32nd Annual Saudi-Japan Symposium on Technology in Petroleum Refining and Petrochemicals 2023

Technology in Fuels & Petrochemicals

Dhahran, Saudi Arabia
5-6 December 2023

ISBN: 978-1-7138-9071-3

Printed from e-media with permission by:

Curran Associates, Inc.
57 Morehouse Lane
Red Hook, NY 12571

Some format issues inherent in the e-media version may also appear in this print version.

Printed with permission by Curran Associates, Inc. (2024)

For permission requests, please contact King Fahd University of Petroleum & Minerals at the address below.

King Fahd University of Petroleum & Minerals
Dhahran 31261
Kingdom of Saudi Arabia

Phone: +966 (13) 860 0000

info@kfupm.edu.sa

Additional copies of this publication are available from:

Curran Associates, Inc.
57 Morehouse Lane
Red Hook, NY 12571 USA
Phone: 845-758-0400
Fax: 845-758-2633
Email: curran@proceedings.com
Web: www.proceedings.com

الندوة السنوية السعودية اليابانية الثانية والثلاثون

32nd Annual Saudi-Japan Symposium

التقنية في الوقود والبتروكيماويات

Technology in Fuels & Petrochemicals

PROCEEDINGS: 32

21-20 جمادى الأولى 1445هـ
December 5-6, 2023

King Fahd University of Petroleum & Minerals
Dhahran, Saudi Arabia

Symposium supported by: KIKX

Preface

This is the Proceedings of the 32nd annual Saudi Japan symposium "*Technology in Fuels & Petrochemicals*" held on December 5-6, 2023 at King Fahd University of Petroleum & Minerals (KFUPM), Dhahran, Saudi Arabia. The event, jointly organized by KFUPM, Saudi Aramco, Japan Petroleum Institute (JPI), and Japan Cooperation Center for Petroleum and Sustainable Energy (JCCP), brings together researchers from industry and academia to discuss recent downstream technologies related to refining and petrochemicals.

The main theme of this year's event is "*Innovative Catalyst Development*" with topics on crude oil to chemicals, methane conversion, using machine learning to accelerate catalyst discovery and predict performance, novel processes for petrochemicals, utilization of refinery waste streams, hydrogen production via direct air electrolysis, sustainability, circularity of plastic wastes, conversion of carbon dioxide to chemicals and jet fuel. There is a total of 22 oral presentations by a distinguished panel of speakers from KFUPM, Saudi Aramco, KAUST, Japanese universities of Hokkaido, Kyushu, Shizuoka, Tohoku, Tottori, Toyama, and Honeywell UOP. In addition, there are 25 poster presentations by young researchers, postdocs and graduate students from Chemistry and Chemical Engineering.

I take this opportunity to thank the President of KFUPM Dr. Muhammad M. Al-Saggaf and the Vice President for Research and Innovation Dr. Ali A. Al-Shaikhi for their continuous support and valuable guidance for holding this Symposium. I would like to thank our counterparts at Saudi Aramco, JPI and JCCP for their efforts throughout the planning stages of the symposium. I also thank all members of our organizing committee and different groups at CRAC for preparing this event.

We extend our thanks to distinguished speakers and attendees for their active participation and look forward to your participation in our next symposium.

Dr. Wael A. Fouad
Director, Center for Refining & Advanced Chemicals
https://ri.kfupm.edu.sa/crac
KFUPM, Dhahran, Saudi Arabia

Contents

Oral 1 (Keynote)

JPI direction along carbon neutrality and introduction of our research, direct methane conversion over zeolite catalysts

Atsushi Muramatsu

International Center for Synchrotron Radiation Innovation Smart, Tohoku University, 2-1-1, Katahira, Aoba-ku, Sendai 980-8579, Japan, Tel & Fax: +81-22-217-5163, E-mail: mura@tohoku.ac.jp

Abstract

Japan has set a target of carbon neutrality by 2050 so that all kinds of technologies and policies will be incorporated. The action plans of the Japan Petroleum Institute (JPI) are proposed to the members. The following are the things we would like to work on together with the JPI members.

1. Proposals backed by science, technology, and theory "Guidelines for building the carbon neutral society and its transition process"
2. Promoting the level up of carbon neutral technology
3. Overseas carbon neutrality trend survey (Europe, the United States, China, oil/gas producing countries, etc.)
4. Collaboration with other academic societies, related organizations, local governments, etc.
5. Public relations and information provision regarding carbon neutrality and oil
6. Providing a place to discuss carbon neutrality

Let's introduce our research on catalysis. Novel synthesis method for zeolite materials has been developed in our laboratory, which is based on mechanochemical reaction of precursor solid oxides, containing SiO_2 and the other metal (hydro)oxides, to give amorphous mixed oxides, and successive hydrothermal synthesis through conversion from them via solution phase. Ce-containing MFI-type zeolite (Ce-MFI)-supported Pd catalysts modified by a second metal were prepared and found active for low-temperature oxidative coupling of methane reactions to give C2 hydrocarbons.

Oral 2

The Origin of Atmospheric Methane in the Eastern Province

Ahmed. M. El Zohry and Theis I. Solling

CIPR, CPG, KFUPM

Abstract

The global consensus is that the world is getting warmer as a result of increased concentrations of greenhouse gasses in the atmosphere. The main contributors are carbon dioxide and methane. Particular focus has been put on the former because of the clear link to anthropogenic activities. More or less any activity that we, as a species, enter into is associated with energy demand which again is supported by burning fossil fuels that lead to the emission of CO_2 into the atmosphere. Since IR (heat) radiation is absorbed by CO_2 a fraction of the heat that overwise would have escaped the planet stays in the system and effectively increases the overall global temperature. The reason behind the absorption is the C=O asymmetric stretch. The absorption efficiency of CO_2 is high in that particular stretch. If one compares CO_2 with for example methane where CH stretching motions can become activated, CH_4 is in fact 36 time more potent as a greenhouse gas. Of course, the sheer volume of CO_2 in the atmosphere "makes up for" the lack of efficiency but the CH_4 concentration has increased much more rapidly compared to CO_2 and is the potential of a much more severe problem going forward. Additionally, as an extra caveat, namely the increased focus on hydrogen. As a reactive gas H_2 has the potential to interfere with the methane balance because H_2 and OH radicals react and leave less OH in the system to clean up residual CH_4 which amplifies the methane global warming potential. Therefore, it is of very high relevance to understand the sources of methane in the overall pool in the atmosphere. Particularly, it is of high relevance in a hydrocarbon-producing country to assess the contribution of biogenic CH_4 versus the CH_4 that leaks from production to eventually reduce the latter.

A study carried out in the Caribbean has shown that in the vicinity of a sea mist iron containing particulate matter is photo-inductively able to convert CH_4 into CO and that the distribution of isotopomers depends on the origin of the CH_4 because of the kinetic isotope effects that are in play during the decomposition. The amount of C-13 varies and is related to the source of the methane. The results of the analysis gave a clear picture of a mostly biogenic source of the methane in the Caribbean. Clearly, that question of the source of the atmospheric methane is even more relevant from the perspective of an oil producing country. The coastline of Saudi Arabia's Eastern Province offers an extremely advantageous venue for exploration. Reason being the particulate matter that blows from the vast red dunes of the empty quarter and highly saline sea spray at the Gulf of Salwa.

The way that the photochemistry works is that the iron particles exchange with the sea spray to form $FeCl_3$. The light is then absorbed to generate excited-state $FeCl_3$ in a sufficiently high concentration to lose Cl radicals or to interact directly with the available methane. The most plausible mechanism is the radical generation with subsequent reaction between Cl and CH_4. This will generate methyl radicals what will react with O_2 to ultimately form various oxygenated species that can be quantified using isotope spectrometers. We have already generated a plethora of results pertaining to the photocatalytic behavior of $FeCl_3$ with an example shown in Figure 1. The most significant absorptions lie well within the attainable solar spectrum.

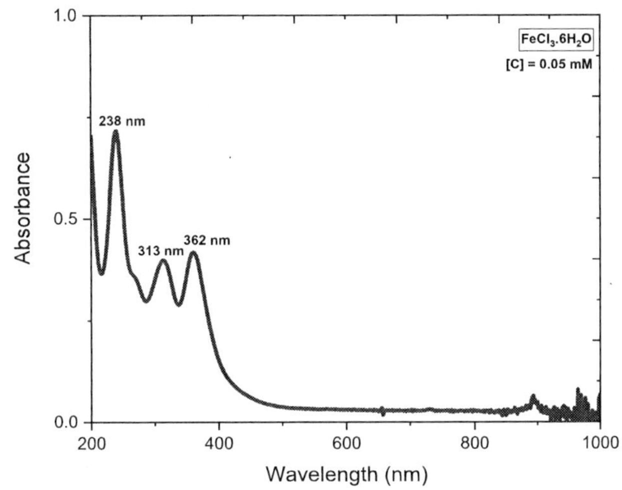

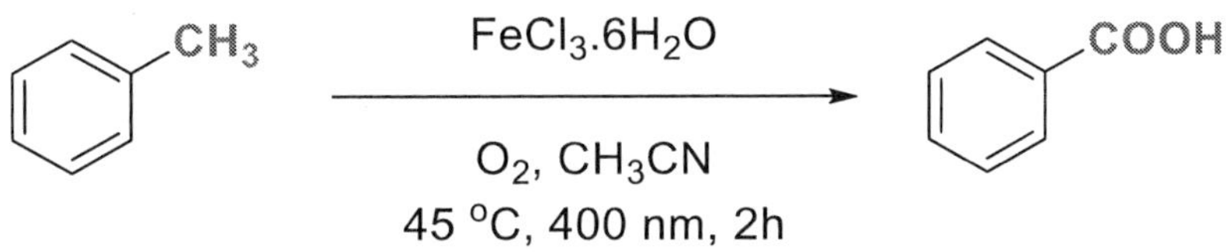

Figure 1: Example of $FeCl_3$ as a plausible photo catalyst for oxidation of hydrocarbons.

At CPG we have established capabilities that allow us to study firstly, ultrafast photo-induced phenomena in our femtosecond laser lab and to conduct state-of-the-art isotope quantification. The femtosecond laser lab evolves around a Ti : Saphire system that gives femtosecond laser pulses at a 5 kHz repetition rate with a pulse power of 5 mJ pulse with a duration of 90 fs. The capabilities here are focused on providing time resolution on the femtosecond time scale along with very high laser fluences. This has assisted us in exploring the photoinduced dynamics of the $FeCl_3$ system. Some results are presented in Figure 2 along with an overview of the laser system.

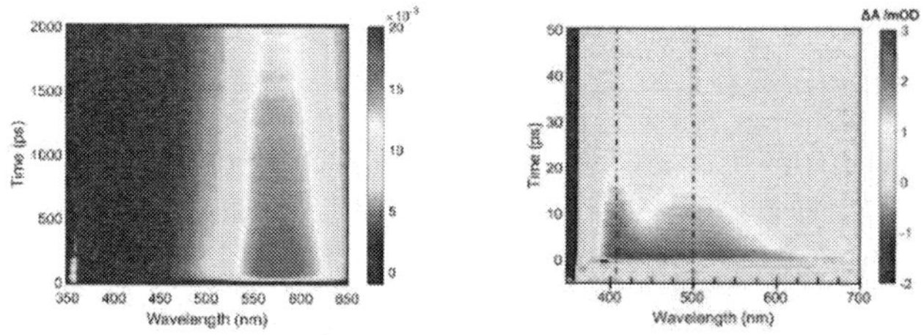

Figure 2: The femtosecond laser lab is shown at the top and two sets of results for the photo-induced dynamics of $FeCl_3$. It can be seen that a state that decays in 2 ps is populated to begin with. Subsequent to its decay two other states that have a lifetime of around 15 ps are populated and are most likely the reactive ones. They stay around long enough for photo chemistry to play out but short enough for the electronic states not to be quenched.

The isotope labs at CPG will provide a detailed isotope analysis. The capacity is there to quantify, to a very high degree of precision, C-12/C-13 and O-16/O-18 ratios. By performing a sampling campaign at the Gulf of Salwa we will measure the $C^{12}O^{16}/C^{13}O^{16}$ and $C^{12}O^{16}/C^{12}O^{18}$ (together with variations thereof). Depending on the deviation (the δ value) from the naturally occurring standard it will be possible to assess whether methane in the Saudi environment arises from biogenic of fossil sources and thereby minimize the concentration in the air and minimize the global warning methane footprint.

Figure 3: The CPG isotope labs are capable of measuring the isotopic distribution in gaseous samples from the Gulf of Salwa.

Oral 3

Crude to Chemicals – Old Challenges and New Opportunities

Lujain R. Alfilfil and Khalid A. Almajnouni

Saudi Aramco

Abstract

The demand on transportation fuels is expected to decrease due to the development of electric, fuel-cell-based and hybrid automobiles while petrochemicals demand will continue to grow globally at a rate of ~ 3-4% (CAGR). The current refinery schemes might need reconfigurations to cope with the new demand. Therefore, the efforts in recent years have been directed to revisiting the product slate from current refineries by changing their operating model and developing an innovative route to favor more petrochemicals production.

In this talk, we will show our efforts on developing a novel process concept for the conversion of untreated Arabian Light to chemicals in a single step using an innovative multi-zone fluidized bed reactor (MZFBR) for maximizing the production of chemicals. A stable continuous operation for more than 8 hours with the production yield of light olefins above 30 % was achieved with this process design. The stripping steam reduced the hard coke formation and the multifunctional catalyst is able to be regenerated in situ using this process.

Oral 4

Application of Machine Learning to Predict Performance of Oxidative Dehydrogenation Catalysts

Gazali Tanimu[1], Akira Yada[2], Jimoh O. Ajadi[1,3], Abdullah M. Aitani[1], Hassan Alasiri[1,4]

[1]*Interdisciplinary Research Center for Refining & Advanced Chemicals, King Fahd University of Petroleum & Minerals, Dhahran, Saudi Arabia; Email address: gazali.tanimu@kfupm.edu.sa*

[2]*Interdisciplinary Research Center for Catalytic Chemistry, National Institute of Advanced Industrial Science and Technology, Tsukuba, Ibaraki 305-8565, Japan.*

[3]*Department of Mathematics, King Fahd University of Petroleum & Minerals.*

[4]*Department of Chemical Engineering, King Fahd University of Petroleum & Minerals*

Abstract

1,3 Butadiene (BD) is an important raw material in the chemical industry utilized in the manufacturing of synthetic rubber and automotive tires. One of the on-purpose BD production methods is oxidative dehydrogenation (ODH) of n-butane. A recently investigated approach targeted at improving the selectivity of ODH catalysts other than the traditional trial-and-error approach involves the utilization of catalyst informatics which applies data science to the field of catalysis. This enables the design of suitable and desirable catalysts from a designated catalytic reaction, catalyst structure, and/or reaction mechanism [1-4]. Catalyst informatics combines knowledge of many areas, such as dataset generation, data preprocessing, data visualization, and machine learning (ML), to speed up the discovery and understanding of heterogeneous catalytic materials. ML utilizes statistical algorithms to identify significant features in datasets and make predictions from these learned relationships between descriptors and measured properties.
Interestingly, the strength of ML algorithms was utilized to predict the performance of catalysts in ODH of n-butane to BD. The datasets were generated from experiments conducted in a fixed-bed reactor in the temperature range of 400-500 ^{o}C and O_2/C_4 molar feed ratio of 1-4 mol/mol, using (0-30 wt %) loading of metal oxides of Ni, Fe, Co, Bi, Mo, Mn and W supported on gamma alumina catalysts. Repeated 10-fold cross-validation was used for training, optimizing and testing the models. Among the 10 different ML models examined, eXtreme Gradient Boosting (XGB) and Gradient Boosted Tree Regression (GBR) models gave the best coefficient of determination (R^2) performance for predicting 1,3-butadiene selectivity (92.0 % & 91.0 %). In addition, the models had the lowest training and test mean absolute errors (MAE: 3.36 & 3.62) and root mean square errors (RMSE: 4.42 & 4.69). The presentation highlights the notable role of ML algorithms in enhancing the prediction of n-butane ODH catalyst performance.

References

[1] K. McCullough, T. Williams, K. Mingle, P. Jamshidi, J. Lauterbach, *Phys. Chem. Chem. Phys.*, **2020**, *22*, 11174-11196.

[2] A. J. Medford, M. R. Kunz, S. M. Ewing, T. Borders, R. Fushimi, *ACS Catal.* **2018**, *8*, 7403-7429

[3] K. Takahashi, L. Takahashi, I. Miyazato, J. Fujima, Y. Tanaka, T. Uno, H. Satoh, K. Ohno, M. Nishida, K. Hirai, J. Ohyama, T. N. Nguyen, S. Nishimura, T. Taniike, *ChemCatChem,* **2019**, *11*, 1146-1152.

[4] S. Mine, Y. Jing, T. Mukaiyama, M. Takao, Z. Maeno, K. Shimizu, I. Takigawa, T. Takao, *Chem. Lett.*, **2022**, *51*, 269-273.

Oral 5

Accelerating Catalyst Discovery Using Extrapolative Machine Learning Approach

Takashi Toyao
Institute for Catalysis, Hokkaido University, Japan

Abstract

Designing novel catalysts is key to solving many energy and environmental challenges. Despite the promise that data science approaches, including machine learning (ML), can accelerate the development of catalysts, truly novel catalysts have rarely been discovered by ML because of one of its most common limitations and criticisms—the assumed inability to extrapolate and identify extraordinary materials. In this context, we have developed a new ML approach that uses elemental features as the input representations rather than using catalyst compositions directly as inputs. This ML method has the potential to guide catalyst design and discovery in areas where there is a limited overlap of the catalyst composition in the training data set. Moreover, this method is applicable even for elements previously never included in the given data, enabling extrapolative and ambitious prediction beyond the training data. In this talk, I will show our recent contributions to developing new heterogeneous catalysts using the extrapolative ML method.

Oral 6

AI and data-driven optimization of one-step Crude to Chemicals process

Noor Sulais, Balaji Mohan, Lujain Filfil, and Khalid Majnouni

Saudi Aramco

Abstract

There is a continuous global demand growth of petrochemicals while oil demand is softening in transportation fuels. Strict Sulphur specifications were recently extended beyond gasoline and diesel to cover other fuels. Thus, refineries have been revisiting their product slate in recent years by changing their operating model to produce more petrochemicals.

After persisted research, a novel process concept to convert untreated Arabian Light to chemicals in a single step was demonstrated using an innovative mutli-zone reactor for maximizing chemicals production. With this process design, more than 30% light olefins were produced in a stable continuous operation of more than 8 hours.

To further enhance the process, this study uses an Artificial intelligence (AI) / machine learning (ML)-based accelerated co-optimization of catalyst composition and reaction conditions to maximize light olefin production. First, a systematic and structured experimental matrix was created using the Latin hypercube space-filling design of experiment (DoE) algorithm. After conducting experiments and obtaining results, an in-house automated SuperLearner (autoSL) tool [1,2] was used to create an ML model for hydrogen transfer and light olefin production using the design matrix as input features. The in-house autoSL tool fits 40 different ML models, selects the top five best-performing models on this data, and builds a SuperLearner model from these top-performing models. Finally, using this SuperLearner model as a surrogate for the experiments, a multi-objective co-optimization was performed to identify the optimal catalyst composition and reaction conditions to minimize hydrogen transfer and maximize light Olefin production through active learning approach.

References

[1] Mohan, B., & Badra, J. (2023). A novel automated SuperLearner using a genetic algorithm-based hyperparameter optimization. Advances in Engineering Software, 175, 103358.

[2] Mohan, B., & Badra, J. (2023). *U.S. Patent Application No. 17/647,375.*

Oral 7

Distillate Hydrocracking Catalyst and Process to Produce Isomerate and Steam Cracking Feedstock

Ashok K. Punetha; Omer R. Koseoglu

Saudi Aramco

Abstract

Hydrocracking processes are used commercially in a large number of petroleum refineries to split the larger molecules of the feed into smaller, i.e., lighter, molecules having higher average volatility and economic value. These processes and techniques have been utilized conventionally in the art for the production of middle distillates. The current market conditions, i.e., expected diesel surplus and growing chemical demand, drives the research to convert diesel either into lighter fuel components or to value-added chemicals. This study was an attempt to develop a catalyst and process that will be a solution to convert surplus diesel to gasoline blending components by dearomatization and hydrocracking reactions.

An ultra-low sulfur diesel received from a refinery was processed in a single reactor pilot plant unit over a catalyst containing both cracking and deep hydrogenation functions. Tests were conducted at hydrogen partial pressure 40, 60, 80 bars, temperature 250–355 °C, at 15 °C increments and H2/oil ratio 1000 NL/L to find the optimum conditions for isomerate production. Products were analyzed for density, SIMDIST, 2D-GCxGC, PIONA and derived cetane number (DCN) analysis to determine the process performance.

Both simulated distillation and density of the products showed increased conversion of diesel into naphtha boiling range hydrocarbons at temperature at 310 ^{0}C and higher. No significant difference was found in product composition at different hydrogen partial pressures but the data indicates that 60 bars of hydrogen partial pressure is needed for stable catalyst life.

The PIONA data of the products obtained at high temperatures revealed that the major portion of compounds are in the range of C_4-C_9, indicating formation of an isomerate fraction, rich in iso-paraffins and fluid catalytic cracking feedstock, rich in naphthenes. The study reveals that the isomerate fraction can be further processed in a reverse isomerization unit to produce normal paraffins, a preferred feedstock for steam cracking.

Oral 8

Molecular Kinetic Model Development and Parameter Estimation for Naphtha Reforming

Syed A. Ali [a], Mohammad M. Hossain [a,b], Rajesh Theravalappil [a], Ali H. Alzaid [c], Ali H. Alshareef [c], Haitao Huang [c]

[a] *Center for Refining & Advanced Chemicals, King Fahd University of Petroleum & Minerals, Dhahran, Saudi Arabia*

[b] *Chemical Engineering Department, King Fahd University of Petroleum & Minerals, Dhahran, Saudi Arabia*

[c] *Research and Development Center, Saudi Aramco, Dhahran, Saudi Arabia*

Abstract

A molecular-level kinetic model of catalytic naphtha reforming is developed using the experimental data generated via well-controlled bench-scale tests conducted at 480-510 °C, 1.5-3.0 h^{-1}, and 3.4 bar over a Pt-Sn/Al_2O_3 catalyst. Thirty-seven chemical species, comprising of 15 pure components and 22 pseudo-components in the C1 to C10 range were formulated to describe the chemical composition of reformate. Differentiation was made in the model between single- and multi-branched *iso*-paraffins as well as between alkylcyclopentanes and alkylcyclohexanes. A set of 181 reactions were carefully selected and incorporated in the kinetic model including dehydrogenation (10), dehydrocyclization (30), hydrodecyclization (30), isomerization of paraffins (30), isomerization of naphthenes (10), dealkylation of aromatics (9), dealkylation of naphthenes (18), hydrocracking of paraffins (41), and ring opening of cyclopentane (3). Reversible reactions are considered as separate forward and backward reactions and catalyst deactivation function was also included. The mathematical model, using MATLAB®, allows estimation of pre-exponential factors for the 181 reactions and activation energies for the nine types of reactions. The estimated rates for various types of reactions were in accordance with the process chemistry. The parity plots show excellent matching of experimental and model predicted data for the seven hydrocarbon groups with an overall coefficient of discrimination (R^2) of 0.9877. Further drilling down of the estimated composition to to molecular-level revealed that the R^2 values for different hydrocarbon types were between 0.872 and 0.994. The molecular kinetic model was validated using a set of data obtained at different temperatures and with another type of feedstock.

Oral 9

Effect of Gallium and Platinum distribution encapsulated in Silicalite-1 (MFI) zeolite on controlled propane dehydrogenation reaction

Fadhil A. Almukhtar

Saudi Aramco

Abstract

The preparation method of the catalyst highly impacts its properties and activity. Optimizing the synthesis conditions mainly targets improving the catalyst performance and overcoming the bottlenecks such as sintering of metal active sites, deactivation, short catalyst lifetime and poor selectivity. In this study, we investigated the influence of the design and preparation method of Silicalite-1 bearing Pt and Ga active species on the properties and the performance of the catalysts for propane dehydrogenation reaction aiming to increase propylene yield. Various synthesis routes, leading to different Pt and Ga location and distances were tested: (1) supporting metals on the zeolite where both Pt and Ga are randomly distributed on the surface, (2) confining of Pt and Ga within the zeolite pores following in-situ approach with no control of their relative positions, and (3) core-shell design where one metal is confined within Silicalite-1 is covered by a Silicalite-1 layer including the second metal. The influence of structure, textural properties, location of Pt and Ga nanoparticles and their synergetic interaction to form Pt-Ga alloys were studied using several characterization techniques such as XRD, BET, TEM-EDX and NMR. Catalytic performance revealed that confining metals improved the selectivity and lifetime of the catalyst. Moreover, spatial separation of Pt and Ga through the core-shell design further boosted the reaction yield with conversions hitting the equilibrium limit. Ga/Pt ratio played a crucial role in tuning the catalyst performance. X%Pt(core)-Y%Ga(shell)@S-1 catalyst with a specific Ga/Pt ratio exhibited superior results of conversion and selectivity.

Oral 10

Highly Efficient Dehydrogenation of Isopentane to Isoprene: Selectivity Control of the Catalytic Reaction Field by Electric Internal Heating System

Ryo Watanabe, Choji Fukuhara

Department of applied Chemistry and Biochemical Engineering, Graduate school of Engineering, Shizuoka university, 3-5-1 Johoku, Naka-ku, Hamamatsu, Shizuoka 432-8561, Japan

Abstract

This study focuses on studying the construction of a dehydrogenation (*iso*-$C_5H_{12} \rightarrow C_5H_8 + 2H_2$) system for isopentane (*iso*-C_5H_{12}) for the production of isoprene, a raw material for synthetic rubber. We have shown that an iron catalyst (Fe/SiO_2) supported on SiO_2 is highly active for the dehydrogenation of *iso*-C_5H_{12}, and that isoprene can be continuously produced in the presence of hydrogen sulfide (H_2S). On the other hand, it has also been shown that an increase in reaction temperature accelerates the gas-phase decomposition of the feedstock and products, resulting in a progressive loss of selectivity to olefins. In this study, an electric internal heating (EIH) catalyst system that selectively heats the catalytic reaction field was developed to improve isoprene yield while suppressing gas-phase pyrolysis. Specifically, Fe/SiO_2 catalyst was coated on a silicon carbide substrate, and *iso*-C_5H_{12} dehydrogenation in the presence of H_2S was performed using it. The results showed that heating of the catalyst part itself was selective in the electric internal heating method compared to the heating method using an electric furnace. Such heating suppressed excessive pyrolysis reactions in the gas phase and resulted in higher selectivities of C5 olefins (isoprene, 2-methyl-1-butene, etc.). The selectivity of C5 olefins was enhanced by increasing the feed gas supply. The increase in feed rate lowered the gas-phase temperature, which was assumed to suppress pyrolysis reactions.

Keywords: Electric internal heating, Dehydrogenation, Selectivity

Oral 11

Insights on CO_2-mediated oxidative dehydrogenation of propane: Aspen plus simulation and *in situ* DRIFT experiment

Yahya Gambo[b], Sagir Adamu[a,b], Rahima A. Lucky[c], Mohammed S. Ba-Shammakh[b], Mohammad. M. Hossain[a,b,]

[a] *Department of Chemical Engineering,* [b] *Center for Refining & Advanced Chemicals, King Fahd University of Petroleum & Minerals, Dhahran, Saudi Arabia*

[c] *Chemical Reactor Engineering Center, Western University, London ON, Canada*

Abstract

With dual prospects of bridging the wide shortfall in the global supply of olefins and promoting CO_2 utilization, CO_2-mediated oxidative dehydrogenation of propane (CO_2-ODHP) as an on-purpose and zero-H_2 input process has gained momentous research attention. Herein, we decoupled the complex reaction network of the CO_2-ODHP process via a rigorous thermodynamic simulation. The simulation revealed crucial catalyst design requirements, and delineated the domain and contribution of individual routes involved. Inspired by the insights, we utilized a facile and scalable solvent-free strategy and developed a novel VOx/Al_2O_3 catalysts with suitable active site diversity as confirmed by XRD, Raman and HRTEM analysis. The optimal catalyst in the investigation was AlV07B, which contained 7 wt.% and 5 wt.% boron and vanadium, respectively, and it exhibited an excellent propylene yield of 34% coupled with less than 10% CO selectivity for 12 hours catalytic operation at 650 °C. This was primarily attributed to the impact of BOx incorporation, where it significantly modulated the surface reducibility and moderated the acidity of the catalyst, as confirmed by H_2-TPR and NH_3-TPD respectively. Also, the coordination environment of the AlV07B catalyst was dominated by less polymerized VOx species (with limited evolution of octahedral V^{5+} species and bulky V_2O_5-like specie), and its band gap energy was significantly higher (2.724 eV) compared to unmodified AlV (2.532 eV), as revealed from UV-vis spectroscopy (UV-vis DRS). It is noteworthy that TGA analysis of the spent AlV07B catalyst showed only a small amount of coke formation and SEM-EDX mapping revealed that the V-site elements remained highly dispersed without observable sintering. Moreover, reaction intermediates evolved at much lower temperature over the AlV07B catalyst as evident from *in situ* DRIFTS experiment. In summary, the thermodynamic simulation revealed that effective kinetic control is indispensable and the synthetic protocol demonstrated in our work opens a new horizon for rational design of CO_2-ODH catalysts via harnessing the distinct catalytic ability of boron and metal oxide-based catalysts simultaneously.

Oral 12

Light Olefins Cracking by Zeolites Prepared from Refinery Waste

Mohammad Reb

Saudi Aramco R&DC, Dhahran, Saudi Arabia

Abstract

Saudi Aramco is investigating the potential use of a refinery waste as a solvent to synthesize zeolites as an active component for industrial catalysts and adsorbents. Utilization of the refinery waste allows Saudi Aramco to reduce processing costs, valorizes the waste-derived solvent and modifies the resulting properties of the produced zeolites, which in turn reduces the zeolite synthesis cost. Understanding the catalytic performance of the zeolites prepared using the waste solvent is the objective of this study, which investigates the conversion of 1-hexene and 2-methyl-2-butene (2M2B) at various process operating conditions over zeolites prepared as a function of different silica-to-alumina ratios (SARs) and refinery waste loading levels.

Six zeolites synthesized using the refinery waste were prepared and characterized in terms of acidity, morphology, and textural properties. The impact of zeolite composition and properties on conversion and selectivity is examined and compared with commercial counterparts with similar SARs. The zeolites synthesized with the refinery waste achieved 80% 1-hexene conversion approaching 60% propylene selectivity, outperforming commercial zeolites (52%). However, 2M2B cracking exhibits slower reaction rates and enhanced oligomerization cracking, resulting in lower conversion (46–61%) and propylene selectivity (22–29%). Notably, the zeolite prepared at a 30% refinery waste loading having a sol-gel SAR of 50 shows the best performance among the all ODSO zeolites in terms of stability and selectivity, with results comparable to the commercial zeolites. Notably, that the refinery waste-based zeolites possess larger crystals and higher acid site density compared with the commercial zeolites leading, generally, to a decreased stability. These findings enhance the understanding of the applicability of utilizing refinery waste-based zeolites in industrial processes and guide the development of improved refinery waste-based catalysts for petrochemical applications.

Oral 13

Carbon Dioxide Hydrogenation to Carbon-Neutral Liquid Fuels by Powerful Catalyst

Noritatsu Tsubaki[1]*, Lisheng Guo[1+]

[1]*Graduate School of Engineering and Sustainable Tech. Center, University of Toyama, Japan;*

tsubaki@eng.u-toyama.ac.jp [+]Now: School of Chemistry and Chemical Engineering, Anhui University, China

Abstract

Long-chain hydrocarbons, as carbon-neutral fuels, have high energy density and facile mobile storage capacity, and have attracted wide attention. Owing to the inertia of CO_2 molecules and imprecise control of carbon chain propagation during CO_2 hydrogenation, efficient catalytic conversion of CO_2 into liquid fuel hydrocarbons remains a huge challenge [1].

Cobalt incorporation in iron-based catalyst is a promising way to realize efficient CO_2 hydrogenation to long-chain hydrocarbons with high-yield production [2]. In our previous report, Co-modified iron catalyst exhibited a low CO selectivity and a high selectivity toward C_{5+} hydrocarbons (57.8%) at a CO_2 conversion of 50.2%, in which the yield of C_{5+} reached 26.7% (**Figure 1a-b**). With Co introduction, the cascade reactions between reverse water gas shift (RWGS) and chain propagation are notably reinforced, and the product distribution deviates from the ASF model via an oxygen-containing intermediate (CO*, HCOO*, CO_3^{2*}, and HCO_3*) pathway (**Figure 1c-d**). Besides, our group also found that ultra-high STY of C_{5+} could be achieved via tailoring the balance between RWGS and chain propagation with the help of multi-promoters strategy [4]. More recently, we first designed microwave-assisted carbon supported iron nanoparticle catalysts for efficiently steering CO_x hydrogenation to heavy hydrocarbons [5-6]. By comparison with ordinary calcination, microwave treatment could significantly shorten the heating time to only few seconds (**Figure 1e**), which sheds a light on the general design of high efficiently supported catalysts for practical industry in future.

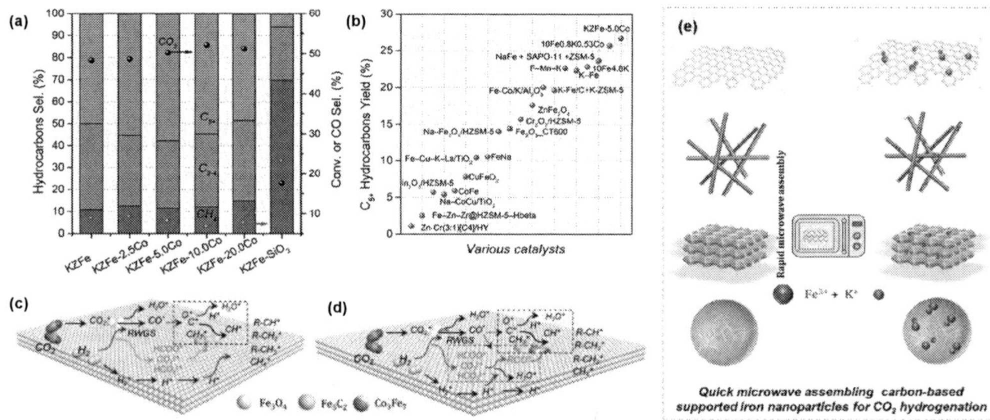

Figure 1. (a) Catalytic performances at 320 °C, 2.0 MPa, and 6000 mL g^{-1} h^{-1}. (b) Yields over various catalysts. (c) KZFe and (d) KZFe-5.0Co, reaction scheme. (e) Synthesis of catalysts via facile microwave heating treatment.

References

[1] L. Guo, X. Guo, Y. He, N. Tsubaki, *J. Mater. Chem. A*, **2023**, *11*, 11637-11669.

[2] L. Guo, N. Tsubaki et al., *Chem. Comm.*, **2020**, *56*, 9372-9375.

[3] L. Guo, N. Tsubaki et al., *Chem. Sci.*, **2023**, *14*, 171-178.

[4] H. Zhao, L. Guo, N. Tsubaki et al., *J. CO2 Util.*, **2021**, *52*, 101700.

[5] L. Guo, N. Tsubaki et al., *Chem. Eng. J.*, **2022**, 429, 132063.

[6] L. Guo, N. Tsubaki et al., *EES Catal.*, **2023**, *1*, 516.

Oral 14

CO_2 Hydrogenation to Lower Olefins Using Iron Supported Catalysts

Mohammed Sanhoob, Huda S. Alghamdi, Afnan M. Ajeebi, M. Nasiruzzaman Shaikh*

*Interdisciplinary Research Center for Hydrogen and Energy Storage (IRC-HES), King Fahd University of Petroleum & Minerals, Dhahran 31261, Saudi Arabia; *Corresponding Author, Phone: +966 13 860 7717; E-mail address: mnshaikh@kfupm.edu.sa (M. Nasiruzzaman Shaikh)*

Extended Abstract

Carbon has played a crucial role in shaping the course of human civilization, especially in the 20th century. The widespread utilization of fossil carbon resources like oil, coal, and natural gas marked the beginning of a transformative era for human society. Nevertheless, the rapid exhaustion of these limited resources has led to a substantial increase in the emission of carbon dioxide (CO_2). The main factor driving this rise originates from burning fossil fuels for diverse energy applications, including generating electricity, producing heat, and powering transportation. The abundant discharge of carbon dioxide (CO_2) has resulted in adverse effects on the environment and human welfare, encompassing phenomena like climate change, the warming of the planet, and acidification of the oceans [1]. While functioning as a greenhouse gas, CO_2 serves as a valuable and renewable source of carbon [2] In response to these challenges and in alignment with objectives to mitigate climate change, there has been a worldwide emphasis on methods and approaches aimed at decreasing atmospheric CO_2 levels, with Carbon Capture and Storage (CCS) playing a prominent role.

Converting CO_2 directly into value-added chemicals not only tackles environmental concerns stemming from excessive emissions but also fosters the resourceful utilization of CO_2 [3]. However, advancements encountered obstacles and reached a stagnation point in the subsequent two decades, primarily due to apprehensions about carbon neutrality. In this context, there were concerns that the energy consumption of the process might counterbalance any potential benefits. A technologically promising approach has arisen in the direct transformation of CO_2 sourced from the atmosphere or industrial activities. This method holds significant potential for achieving carbon neutrality, particularly when driven by renewable energy within a closed carbon cycle [4]. Aligned with the principle of sustainability and the continuous progress in the utilization of renewable energy, the process of hydrogenating CO_2, employing hydrogen and electricity generated from solar and wind energy, has experienced a resurgence in research interest since 2010.

Several catalytic reduction techniques, encompassing thermocatalysis, photocatalysis, electrocatalysis, and multi-catalytic strategies, facilitate the transformation of CO_2 into a range of platform chemicals. These include methane (CH_4), carbon monoxide (CO), methanol (CH_3OH), formic acid, higher alcohols and olefins [5]. The hydrogenation of CO_2 utilizing heterogenous catalysts possess several advantages such as improved catalyst recycling, stability, and optimized reactor design in contrast to the homogeneous catalytic hydrogenation of CO_2 [6].

The intricate process of converting CO_2 into olefins is accomplished using either the methanol-mediated (MeOH-mediated) route or the Reverse Water-Gas Shift - Fischer-Tropsch Olefins (RWGS-FTO) route, with the employment of appropriate catalysts. Metal-zeolite catalysts are utilized for the methanol-mediated pathway, while Fe-based or Co-based catalysts are employed in the RWGS-FTO route [7]. These routes represent innovative strategies for transforming carbon dioxide (CO_2) into olefins. In the methanol-mediated pathway, CO_2 is initially converted into methanol (CH_3OH), a versatile chemical precursor. This methanol can then undergo dehydration processes to yield valuable olefins, such as ethylene and propylene. On the other hand, the RWGS-FTO route involves a sequential two-step process. In the first step, CO_2 is subjected to the reverse water-gas shift (RWGS) reaction, producing carbon monoxide (CO). Subsequently, the obtained CO is utilized in Fischer-Tropsch Olefins (FTO) synthesis, resulting in the generation of olefins. Substantial progress has been achieved in advancing individual stages related to the development of catalysts and comprehending the mechanisms underlying reactions. Nevertheless, there is a current shift in focus towards facilitating the direct hydrogenation of CO_2 to olefins using integrated tandem catalysts, marking a novel direction in research [8-10].

In a recent study, Lee et al. developed $CoFe_2O_4$/CNT catalyst with Na as a promoter [11]. The results elucidated that they achieved approximately 34% CO_2 conversion and a remarkable 39% selectivity for light olefins. The performance of Na- $CoFe_2O_4$/CNT was quite remarkable when compared to single-metal nanoparticle catalysts (Na-Fe_2O_4/CNT and Na-Co/CNT) and their physical mixture ((Na-Fe_2O_4 + Co/CNT). The outstanding performance was ascribed to the facile generation of a distinctive bimetallic alloy carbide active species, precisely identified as $(Fe_{1-x}Co_x)_5C_2$. Recently, Li et al. presented a novel method for synthesizing bio-ZSM-5, characterized by gentle acid strength and comparatively reduced acid density, making use of natural rice husk as a template [12]. Although the surface area of prepared Bio-ZSM-5 was lower in comparison to commercially available ZSM-5, the ZnZrOx and bio-ZSM-5 catalysts but its catalytic performance was similar to ZnZrOx and ZSM-5. The silica post-treatment of ZnZrOx and bio-ZSM-5 resulted the enhancement in intracrystalline mesopores and acidity. At 380 °C and 3 MPa, the adapted catalyst demonstrated a selectivity of 64% for C2-C4 compounds and less than 1% for aromatics. The introduction of zinc into the ZnZr oxide phase enhanced the creation of oxygen vacancies, actively influencing CO_2 hydrogenation via the methanol-mediated route involving formate.

Therefore, as a part of our continuous effort to develop heterogeneous catalysts, we have developed a series of K- K-promoted iron oxide-containing zirconium oxide composite catalysts for the carbon dioxide hydrogenation reaction. Here, 30wt% of iron oxide on zirconium oxide composite catalyst produced the best results in terms of conversion (22%) with the light olefin selectivity of 35% under 30 bar hydrogen pressure at 375 C with the GHSV of $4500 ml.g^{-1}.h^{-1}$. The role of the concentration of promoters and Fe contents were investigated, and their effect on light olefin selectivity was evaluated.

References

[1] Porosoff, M. D., Yan, B., & Chen, J. G. (2016). Catalytic reduction of CO_2 by H_2 for synthesis of CO, methanol and hydrocarbons: challenges and opportunities. Energy & Environmental Science, 9(1), 62-73.

[2] Matthews, H. D., Zickfeld, K., Knutti, R., & Allen, M. R. (2018). Focus on cumulative emissions, global carbon budgets and the implications for climate mitigation targets. Environmental Research Letters, 13(1), 010201.

[3] Kattel, S., Liu, P., & Chen, J. G. (2017). Tuning selectivity of CO_2 hydrogenation reactions at the metal/oxide interface. Journal of the American Chemical Society, 139(29), 9739-9754.

[4] Wang, D., Xie, Z., Porosoff, M. D., & Chen, J. G. (2021). Recent advances in carbon dioxide hydrogenation to produce olefins and aromatics. Chem, 7(9), 2277-2311.

[5] Zhang, X., Yuan, C., Huang, Z., Xu, H., & Shen, W. (2023). Recent advances in catalytic hydrogenation of CO_2 into value-added chemicals over oxide–zeolite bifunctional catalysts. Catalysis Today, 114219.

[6] Yang, X., Su, X., Chen, D., Zhang, T., & Huang, Y. (2020). Direct conversion of syngas to aromatics: A review of recent studies. Chinese Journal of Catalysis, 41(4), 561-573.

[7] Gambo, Y., Adamu, S., Lucky, R. A., Ba-Shammakh, M. S., & Hossain, M. M. (2022). Tandem catalysis: A sustainable alternative for direct hydrogenation of CO_2 to light olefins. Applied Catalysis A: General, 641, 118658.

[8] Ye, R. P., Ding, J., Gong, W., Argyle, M. D., Zhong, Q., Wang, Y., ... & Yao, Y. G. (2019). CO_2 hydrogenation to high-value products via heterogeneous catalysis. Nature communications, 10(1), 5698.

[9] Zhou, W., Cheng, K., Kang, J., Zhou, C., Subramanian, V., Zhang, Q., & Wang, Y. (2019). New horizon in C1 chemistry: breaking the selectivity limitation in transformation of syngas and hydrogenation of CO_2 into hydrocarbon chemicals and fuels. Chemical Society Reviews, 48(12), 3193-3228.

[10] Ojelade, O. A., & Zaman, S. F. (2021). A review on CO_2 hydrogenation to lower olefins: Understanding the structure-property relationships in heterogeneous catalytic systems. Journal of CO2 Utilization, 47, 101506.

[11] Kim, K. Y., Lee, H., Noh, W. Y., Shin, J., Han, S. J., Kim, S. K., ... & Lee, J. S. (2020). Cobalt ferrite nanoparticles to form a catalytic Co–Fe alloy carbide phase for selective CO_2 hydrogenation to light olefins. ACS Catalysis, 10(15), 8660-8671.

[12] Li, W., Wang, K., Zhan, G., Huang, J., & Li, Q. (2021). Design and Synthesis of Bioinspired ZnZrOx &Bio-ZSM-5 Integrated Nanocatalysts to Boost CO_2 Hydrogenation to Light Olefins. ACS Sustainable Chemistry & Engineering, 9(18), 6446-6458.

Oral 15

Insights into Plastic Pyrolysis and Bio-Oil Upgrading: Process Optimization and Techno-Economic Analysis

Omar Abdelaziz[1,2,*], Tove Kristensen[3,4], Simon Dahl[3], Christian Hulteberg[3,4]

[1]*Department of Chemical Engineering, King Fahd University of Petroleum & Minerals, Dhahran, Saudi Arabia;* *omar.abdelaziz@kfupm.edu.sa

[2]*Interdisciplinary Research Center for Refining & Advanced Chemicals, King Fahd University of Petroleum & Minerals, Dhahran 31261, Saudi Arabia*

[3]*Department of Chemical Engineering, Lund University, 221 00 Lund, Sweden*

[4]*Hulteberg Chemistry & Engineering AB, 212 25 Malmö, Sweden*

Abstract

The need for plastics is increasing at a rapid pace due to the extensive applications of plastics in various sectors, and recycling infrastructure must be improved to make plastic consumption more sustainable. The conversion of plastic waste into liquid bio-oil via pyrolysis has an immense potential as a chemical recycling approach, where the bio-oil generated has properties comparable with commercial fuels. The fundamental understanding of the various process parameters influencing the upgrading of pyrolytic bio-oils is also of significance. These aspects are key for enabling further progression and improvement in the industrial plastic conversion and bio-oil upgrading methods. In this presentation, an innovative demonstration of the effects of operating conditions on the continuous hydrodeoxygenation of a bio-oil model over a commercial Ni–Mo/δ-Al_2O_3 catalyst is discussed. Response surface methodology was used for the statistical design of experiments, and the application of central composite design enabled identifying the optimal operating conditions with respect to temperature, pressure, and space velocity. Modeling and techno-economic analysis of plastic pyrolysis was performed, and products from the model included 20% of liquefied petroleum gas and 28% of fuels such as gasoline, kerosene, and diesel based on a pre-sorted polyethylene feed. Process integration opportunities into existing petroleum refinery infrastructures was also analyzed.

Oral 16

Selective Aromatics Recovery by Catalytic Conversion of Pyrolysis Gas from Carbon Fiber Reinforced Plastic

Kazumasa Oshima

Department of Chemical Engineering, Kyushu University, 744, Motooka, Nishi-ku, Fukuoka 819-0395, Japan, Tel: +81-92-802-2762, E-mail: koshima@chem-eng.kyushu-u.ac.jp

Abstract

The global push for material circulation has escalated recently, with a particular focus on feedstock recycling of waste plastic as a valuable carbon resource. Carbon fiber reinforced plastics (CFRP), a composite material of carbon fibers and resin such as epoxy, have attractive properties such as lightweight and high strength and are widely utilized for airplanes and wind blades. Because CFRP is generally recycled in a pyrolysis method at low temperatures to avoid thermal damage on carbon fibers, various heavy hydrocarbons are generated due to incomplete decomposition. This study investigated the catalytic conversion of hydrocarbons generated in CFRP pyrolysis to form valuables using zeolite catalysts to promote CFRP recycling. The activity tests were performed using a tandem reactor to separately evaluate the CFRP pyrolysis and catalytic conversion. The result showed high phenol selectivity via the catalytic conversion of the CFRP pyrolysis gas. The advantage of the process is that there is no additional heat supply for the catalytic conversion. In addition, suitable conditions for high phenol selectivity were found through investigation of the reaction mechanism. The high selectivity is not achievable in mixed pyrolysis, suggesting the possibility of a new plastic recycling system as materials circulation.

Oral 17

Challenges and Opportunities in Converting Waste Plastic into Value-added Products

Mohammad Nahid Siddiqui

Chemistry Department and IRC for Refining and Advanced Chemicals, King Fahd University of Petroleum & Minerals (KFUPM), Dhahran, Saudi Arabia, E-mail: mnahid@kfupm.edu.sa

Abstract

Plastics are used in a wide variety of products and have displaced other materials -such as wood, metal, and glass due to the ease of availability of raw materials and cost-effectiveness. Packaging is one of the main applications of plastic along with bottles. The worldwide production of plastics is consistently increasing and has reached an overwhelming 390.7 million metric tons in 2021 an increase of 4% from the year 2020. Asia is the largest plastics producer in the world and China alone accounted for 32% of global production in 2021. In recent years, China produced between 6 and 12 million metric tons of plastic products each month. North America ranks second worldwide with a production of 18% in 2021. The report predicted that the production capacity of the Gulf plastic industry would grow to 34.5 million tons by 2022, at an annual growth rate of 3%. Saudi Arabia has more than 1,321 plastics factories with over SAR 35 billion in investments. Saudi Arabia produces about 53 million tons of waste every year, contaminating its soil and groundwater. According to Saudi Arabia's National Center for Waste Management, the environmental degradation caused by solid waste in 2021 was estimated at $1.3 billion. With investments totaling SR35 billion ($9.3 billion) and with over 1300 factories involved, Saudi Arabia's plastic material and manufacturing has come front and center in the kingdom.

In recent times, the Kingdom has taken a giant leap towards a circular economy where converting waste plastics into value-added products plays a significant place. There are several options for recycling waste plastics but Chemical recycling has been opted as the most suitable for the economic recovery. Chemical recycling involves the breakdown of the plastics into their original monomers or different value-added products. We have conducted several research studies on converting polyethylene, PS, PET, PP, and PVC into useful products and some of the data will be presented. Some future work will also be discussed.

Oral 18

Challenge of Green Hydrogen Production from Air by Direct Air Electrolysis

Etsushi Tsuji

Center for Research on Green Sustainable Chemistry, Faculty of Engineering, Tottori University, Japan

Abstract

Green hydrogen, which is formed by water electrolysis using renewable energies such as solar and wind powers, has attracted much attention from the points of view of clean energy resources, energy carriers and chemical feed stocks. Water electrolysis usually use freshwater, however, it is difficult to even access to freshwater for daily life in many parts of the areas rich of the renewable energies, for example drylands in the Middle East, North Africa, southwest of North America and so on. On the other hand, even in the Sahel desert, the average relative humidity is ~20%. Therefore, more recently, we have started to study about water electrolysis using in-situ captured water from air, "Direct Air Electrolysis (DAE)". In our DAE system, NaClO4, which is a deliquescent neutral salt, was used as electrolyte of water electrolysis because it can absorb water vapor from air and was not decomposed during water electrolysis. Current density of water splitting using the DAE was observed after leaving of the DEA cell for ~10 hours in humidified air, and that after 20 hours was comparable to that using a normal water electrolysis system.

Oral 19

Adjusting the crude oil–to–chemical process using unconventional reactors and catalyst formulations

Pedro Castano

KAUST Catalysis Center (KCC) | King Abdullah University of Science and Technology (KAUST), Thuwal, 23955-6900 Saudi Arabia

Abstract

Crude oil catalytic cracking to petrochemicals is an exciting avenue to guarantee refinery profits, yet it has significant problems. We investigate several reactor and catalyst formulation modifications to improve the conversion of crude and the yield of chemicals. These modifications include using multifunctional reactors (multizone fluidized bed reactor) and adding silicon carbide to the catalyst formulation. We demonstrate that with this bottom-up approach, we can control the morphology of the catalyst, the fluid dynamics, the heat transfer, and, ultimately, the light olefin yield.

Oral 20

Fueling Future Aviation with CO_2 to Jet Technology

Arthur Foutsitzis

Honeywell UOP, KSA

Abstract

As the world strives to reduce greenhouse gas emissions, Sustainable Aviation Fuel (SAF) will continue to be a rapidly growing market. The available global triglyceride feedstocks are insufficient to meet the increasing demand for SAF and additional pathways are necessary. Meeting future demand will require efficient "power to liquids" pathways, which produce e-Fuels from CO_2 using green electricity. This presentation provides an overview of what drives SAF demand, the most promising routes to SAF today, and the UOP methanol to jet process.

Oral 21

3D Porous Polymers for Selective Removal of CO_2 and H_2 Storage

Othman Sadeq Al Hamouz

Chemistry Department, King Fahd University of Petroleum & Minerals, Dhahran

Abstract

In this article, newly designed 3D porous polymers with tuned porosity were synthesized by the polycondensation of Tetrakis(4-aminophenyl)methane with pyrrole to form **M1** polymer, and with phenazine to form **M2**. The polymerization reaction used *p*-Formaldehyde as a linker and nitric acid as a catalyst. The newly designed 3D porous polymers showed permanent porosity with a BET surface area of 575 m^2/g for **M1** and 389 m^2/g for **M2**. The structure and thermal stability were investigated by solid ^{13}C-NMR, FT-IR, and TGA. The performance of the synthesized polymers toward CO_2 and H_2 was evaluated, demonstrating adsorption capacities of 1.85 mmol/g and 2.10 mmol/g for CO_2 by **M1** and **M2**, respectively. The importance of the synthesized polymers lies in their selectivity for CO_2 capture with CO_2/N_2 selectivity of 43 and 51 for **M1** and **M2**, respectively. **M1** and **M2** polymers showed their capability for hydrogen storage with a capacity of 66 cm^3/g (0.6 wt.%) and 87 cm^3/g (0.8 wt.%), respectively at 1 bar and 77 K. The exceptional removal capabilities combined with the high thermal stability and microporosity enables **M2** to be a potential material for flue gas purification, and for hydrogen storage.

Oral 22

Metal-Organic Frameworks Functionalization and Design Strategies for CO_2 Capture

Nawal M. Alghoraibi

Saudi Aramco R&DC

Abstract

Carbon capture from the atmosphere and fixed-point sources is an effective venue to offset CO_2 emissions to achieve reduction targets. Current metal hydroxide or amine-based technologies operate through selective chemical reactions with CO_2 and on thermal cycles; however, low separation capacity, high cost, and major energy losses due to regeneration has limited the technology from reaching wide-range commercial scale.

Metal-organic frameworks (MOFs) are a class of porous materials that have attracted great attention in gas separation and adsorption over the past years due its flexibility, porosity, capacity, stability, and surface functionality. Despite the excellent performance of MOFs in the application of direct air capture (DAC) compared to other technologies, challenges in capture capacity, selectivity, and cost due to the low concentration of CO_2 still remain in this area which showcase the need for further development and material optimization.

In this work, we present chemical approaches to enhance the CO2 capture capacity, selectivity, and stability of MOFs to tackle some of the challenges of CO2 capture technologies. Tailoring the surface morphology, porosity, surface area, and hydrophobicity of MOFs through chemical functionalization has shown promising results that greatly enhanced the performance of MOFs. Increasing the CO2 capture capacity and resistance to trace impurities carry the potential to further reduce the CO2 capture process CAPEX and OPEX.

Poster 1

2-Pentene cracking over bifunctional MFI-type zeolites

Mohammad Alkhunaizi, Munir Khokhar, Mohammad Abdullah
Saudi Aramco, Dhahran, Saudi Arabia

Ziyauddin Qureshi, Yahya Gambo, Mohammed Siddiqui, Abdullah Aitani
Center for Refining & Advanced Chemicals, King Fahd University of Petroleum & Minerals, Dhahran, Saudi Arabia

Abstract

Propylene and ethylene are key building blocks for the petrochemical industry whose demand has been increasing over the recent years. One route to produce these light olefins is through catalytic cracking of C4 – C6 olefins rich stream. In this study, a 2-pentene model compound was converted over a modified ZSM-5, impregnated with phosphorous of different concentrations and screened at different WHSVs to measure their impact on the overall reaction performance. The prepared catalysts were also characterised using different characterisation techniques and benchmarked to the parent ZSM-5. The conclusions drawn from this work can be used to study real-life olefinic streams available in the refinery, and set the basis for a thorough kinetic study in the future.

Poster 2

Post-Synthesis Functionalization of Covalent Organic Frameworks for Carbon Dioxide Capture from Air (DAC)

Mona Al-Otaibi, Mahmoud Abdelnaby, Aasif Helal, Bosirul Haque, Ammar AlAhmed

Saudi Aramco, Dhahran, Saudi Arabia

Abstract

Carbon dioxide direct air capture (DAC) is a crucial part of the attempts to manage climate change. DAC requires a highly selective system to sort out an extremely low concentration of CO_2 (~400 ppm) among other dominant components (N_2, O_2) in the air. A substance needing reactive CO_2-specific site is necessary to extract CO_2 from such a diluted source. Aqueous solutions of mono-ethanolamine have already been used for this separation. However, using them results in high energy consumption and has operational issues, including decomposition and toxicity.

In order to minimize the energy consumption compared to aqueous amine solutions, many classes of solid sorbents have been explored for this purpose. ARAMCO collaborates with KFUPM to develop porous material termed covalent organic frameworks (COFs) and its functionalization with aliphatic amines. Here, we present the utilization of COFs as solid-state platforms for the covalent bonding of amines that can be employed to capture CO_2 from the air. Utilizing COFs has advantages for finely tuning framework design and modifying the chemical environment within their pores. Currently, olefin and dioxane-based COFs, namely V-COF-1 and COF-316, have been synthesized and modified with amine species to improve their performance towards selective capture of CO_2 gas. The chemistry of framework conversion and corroborated the accelerated CO_2 uptake phenomena was confirmed through the Fourier transform infrared spectroscopy (FT-IR) and solid-state nuclear magnetic resonance spectroscopy. Interestingly, these aminated platforms showed an enhancement of CO_2 by 40% to the original materials.

Key Words: *Covalent Organic Frameworks, CO_2 Capture, Direct Air Capture, Amine Functionalization, Adsorption.*

Poster 3
Resilience of Transalkylation Catalyst towards Industrial Contaminants

Veera Venkata RK Tammana, Mosab T. Kheyami, Ali N. Jishi, Mohammad I. AlAmer
Saudi Aramco, Dhahran, Saudi Arabia

Syed Ahmed Ali, Ziyauddin Qureshi
Center for Refining & Advanced Chemicals, King Fahd University of Petroleum & Minerals, Dhahran, Saudi Arabia

Abstract

Conversion of heavy reformate with toluene feed to produce mixture of xylenes are typically done over a metal-loaded transalkylation catalyst in pressurized H_2 environment. A series of possible contaminants in typical industrial settings such as 1,2,4,5-tetramethylbenzene (1,2,4,5-TeMB), indene H_2S and CO were tabulated and then tested on a bench-top reactor system using optimized process conditions for this novel transalkylation catalysts. These contaminants were tested, by studying the effect on catalyst performance, and the reversibility of the liquid and gas contaminants was analyzed. The presence of more than a critical amount of water in the reformate feed caused significant reduction in the dealkylation of methyl-ethyl-benzenes (MEBs) and formation of xylenes due to competitive adsorption of water resulting in a decrease in catalyst's acidity. The results obtained and the performance analysis of this catalyst against the listed contaminants and role of the water in the feed will be discussed in the presentation.

Poster 4

Polypropylene Cracking Proceeding in Micropores of MFI Type Zeolite

Tomohiro Fukumasa

Center for Research on Green Sustainable Chemistry, Faculty of Engineering, Tottori University

Abstract

Polypropylene (PP) is converted into liquid hydrocarbons enriched with monocyclic aromatics and C5-10 aliphatics with a solid acid catalyst at about 673 K without pyrolysis process. Although the use of solvent is necessary in this process, normally solvent is cracked and consumed with plastics. Therefore, we found the way which bulky cyclooctane solvent is recovered and only PP is cracked by applying shape selectivity of zeolite. This result indicates that cyclooctane does not react unless enter inside the micropore of zeolite, but the location where PP reacts is not clear. In this study silica-base treatment at small amount of NaOH increased the number of external surface acid sites due to the desilication and exposing of internal acid sites, large amount of NaOH further selectively dissolved the region consisting of SiO_2 of zeolite but the region around Al was not dissolved, resulting in the forming of wormhole-like mesopores and increasing of external surface area. The catalytic activity of PP cracking was almost unchanged at small amount of NaOH and increased at large amount of NaOH similarly to the change of external surface area. The increase of external surface area means the increase of micropore entrances. Therefore, it is considered that PP mainly reacts in the micropore, and in more detail, in the region near to the entrance in micropore.

Poster 5

Towards Sustainable CO_2 Valorization: Harnessing Cu Single Atoms and Nanoparticles

Esraa Kotob[a], Saheed A. Ganiyu[a,b], Mustapha Umar[a] and Khalid Alhooshani[a,b]

[a] *Department of Chemistry, King Fahd University of Petroleum & Minerals, Dhahran, Saudi Arabia.*

[b] *Center for Refining & Advanced Chemicals, King Fahd University of Petroleum & Minerals, Dhahran, Saudi Arabia*

Abstract

The electrochemical transformation of CO_2 into valuable multi-carbon substances offers an encouraging route to leverage renewable electricity in tackling the issue of anthropogenic carbon emissions. However, the limited selectivity of existing catalytic systems in producing multi-carbon products represents a considerable challenge to the broad and effective adoption of this technology. In this study, we unlock the issue of selectivity by mixing recipe of Cu nanoparticles with $CuNO_3$ to obtain Cu-Cu single atom and nano-particles (SANP) catalyst. We observed that copper (Cu) nanoparticles effectively promoted the generation of carbon monoxide (CO) as the first step, while Cu-Cu single-atom mixture played a pivotal role in accelerating the kinetics of carbon-carbon (C-C) coupling reactions. As a result, an impressive Faradaic efficiency (FE) for the production of multi-carbon compounds was achieved.

Poster 6

Catalytic Conversion of LDPE Plastic via Pyrolysis Process

Feras Alqudayri, M. N. Akhtar, and N. Ahmad

Center for Refining and Advanced Chemicals, King Fahd University of Petroleum & Minerals, Dhahran, Saudi Arabia

Abstract

The production of plastic material is increasing with increase in human population because of their excellent properties such as natural plasticity, chemical stability, insulation etc. The annual production of plastic material is ~400 Mt and is expected to reach 1150 Mt by 2050. The plastics waste generally do not degrade under natural conditions, their improper disposal cause a series of environmental problems. The major portion (~80%) of plastic waste is disposed in Landfill causing environmental problems. The catalytic pyrolysis emerges as a promising technology capable of converting waste polymers into high value chemical products, such as olefins, Aromatics etc.

In this study catalytic pyrolysis of Low-Density-Polyethylene (LDPE) was performed using a semi batch reactor. The LDPE mixed with catalyst was loaded in the reactor and was purged with argon to remove the air. The reactor was equipped with mechanical stirrer and was heated to a desired temperature using controlled heating rate. Several zeolites (Zeolite-β, Zeolite-Y, ZSM-5, and Mordenite) having different physico-chemical properties were evaluated for catalytic pyrolysis of LDPE at 350 °C using different polymer to catalyst ratios. The products were collected in the form of liquid and gas after passing through gas/liquid separator. The solid material left over after pyrolysis was collected and weighed. The conversion of LDPE was calculated using following formula:

$$\text{Conversion (\%)} = \frac{(\text{Mass of LDPE Feed} - \text{Mass of solid Residue})}{\text{Mass of LDPE Feed}} \times 100$$

The findings underscore that the type of catalyst and pyrolysis conditions significantly influence the yield and composition of the pyrolysis products. As illustrated in Figure 1, among all the tested catalysts, the ZSM-5 (SiO_2/Al_2O_3: 30) catalyst demonstrated the highest conversion (100%) and liquid products (56%). This optimal outcome was achieved after testing multiple catalyst-to-plastic ratios, with the ratio of 1:20 at a pyrolysis temperature of 350°C proving to be the most effective. The findings of this study suggest that catalytic pyrolysis can be a viable technology for producing BTEX fractions from waste polymers, which can help to reduce plastic waste pollution and conserve resources.

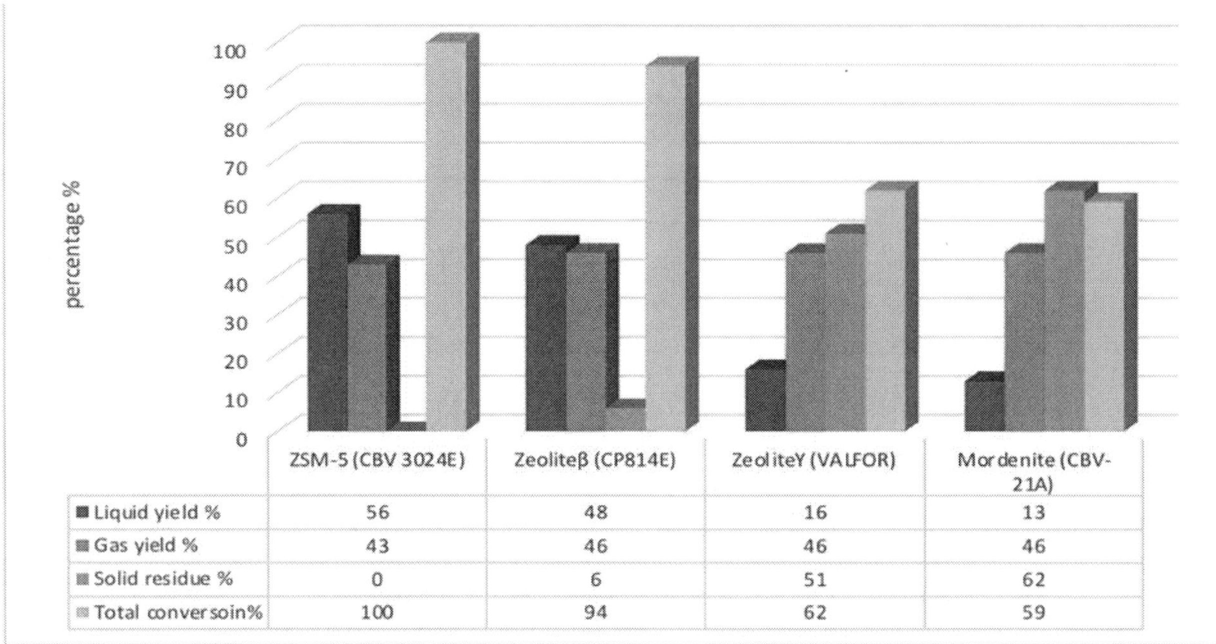

Figure 1: Conversion of LDPE during catalytic pyrolysis over different zeolites.

Poster 7

Thermocatalytic Pyrolysis of Microalgae Biomass over Y-zeolite Catalyst towards Clean Fuel and Valuable Chemicals

Hayat A. Haddad[b], Ahmad Nawaz[a] and Shaikh A. Razzak[a,b]

[a] *Center for Refining & Advanced Chemicals, King Fahd University of Petroleum & Minerals, Dhahran, Saudi Arabia*

[b] *Department of Chemical Engineering, King Fahd University of Petroleum & Minerals, Dhahran, Saudi Arabia*

Abstract

The quest for sustainable energy alternatives has led to the exploration of innovative methods for biofuel production. In this study, the potential of Chlorella vulgaris (CV) biomass as a feedstock for biofuel generation was investigated through thermocatalytic pyrolysis in a cylindrical pyrolizer. Utilizing the response surface methodology (RSM), critical parameters, including temperature, heating rate, and nitrogen flow rate, were optimized to enhance the pyrolysis process. To further understand the influence of catalysts on the pyrolysis outcome, the biomass was subjected to varying catalyst-to-biomass (C/B) ratios, (1:3, 1:6, 1:9, and 1:12). Notably, standard pyrolysis yielded a bio-oil content of 45.67 wt. %, which was higher than the 41.71 wt. % obtained from catalytic pyrolysis at the 1:9 C/B ratio. However, the introduction of the catalyst, while reducing the bio-oil yield, significantly enhanced the quality of the resulting fuel. Characterization of the bio-oils using Fourier-transform infrared spectroscopy (FTIR) highlighted the presence of varied functional groups. Moreover, Gas Chromatography-Mass Spectrometry (GCMS) analysis under optimized conditions revealed an enriched hydrocarbon composition in the catalytic bio-oil, indicating its superior fuel properties. Concurrently, the biochar produced exhibited characteristics that make it a promising material for diverse industrial and agricultural applications. This research underscores the efficacy of thermocatalytic pyrolysis of CV biomass, emphasizing its potential as a renewable source for high-quality biofuel and valuable byproducts.

Keywords: Pyrolysis; Microalgae; Catalyst; RSM; Bio-oil

Poster 8

Effect of plastic composition on the synergetic interactions, kinetic and thermodynamic properties from the copyrolysis of date palm waste and waste foam

Ahmad Nawaz[a,b] and Shaikh A. Razzak[a,c]

[a] *Center for Refining & Advanced Chemicals, King Fahd University of Petroleum & Minerals, Dhahran, 31261, Saudi Arabia*

[b] *Department of Chemical Engineering & Technology, Indian Institute of Technology (BHU),*

Varanasi-221005, India

[c] *Department of Chemical Engineering, King Fahd University of Petroleum & Minerals, Dhahran, Saudi Arabia*

E-mail: ahmadnawaz.rs.che17@itbhu.ac.in

Abstract

Current ways of disposing of plastic garbage, such as dumping or burning, pose enormous ecological problems while also causing irrevocable harm to precious assets. This is especially true for constituents with complicated structures, such as foams. The use of waste materials to manufacture alternative fuels is especially intriguing since co-pyrolysis of waste plastics and biomass has recently been suggested as a potential approach because the synergistic impact may improve product qualities compared to those produced by separate pyrolysis. In context of this, the present study focused on the effect of plastic composition on the synergistic interactions, kinetic and thermodynamic properties from the copyrolysis of date palm waste (DP) and foams. Co-pyrolysis was performed in a thermogravimetric analyser at three different heating rates (10, 20, and 30 °C/min) and at varying biomass to plastic ratios. The kinetic parameters were evalated using Kissinger Akahira Sunose (KAS), Vyazovkin (VZK), and Ozawa Flynn Wall (OFW) isoconversional techniques, and the reaction mechanism was determined using the Z(α) master plot approach. The results of this study suggest that incorporating waste date palm with foams might be a viable option that would result in improved process development.

Keywords: Co-pyrolysis, Foams, Date palm, Kinetics, Thermodynamics

Poster 9

Enhanced selectivity of benzene-toluene-ethyl benzene and xylene (BTEX) in direct conversion of n-butanol to aromatics over Zn modified HZSM5 catalysts

Tatinaidu Kella [a] and Debaprasad Shee [b]

[a] *Center for Refining and Advanced Chemicals, King Fahd University of Petroleum and Minerals, Dhahran, Saudi Arabia*

[b] *Department of Chemical Engineering, Indian Institute of Technology Hyderabad, Kandi, Sangareddy, Telangana, India*

Abstract

BTEX aromatics (benzene, toluene, ethyl benzene, xylenes) are considered as important building blocks chemicals to produce commodity chemicals. Fossil fuels-based feedstocks (petroleum, coal, natural gas) are the main source for the production of these petrochemical intermediate feedstocks. Growing environmental issues, climate change, and depletion of fossil fuels are the motivation to explore the renewable feedstock for the production of fuels and chemicals. The selective production of aromatics from biomass derived alcohols has several applications in the petrochemical, polymer, pharmaceutical industries etc.

Direct conversion of n-butanol to aromatics using xZn-HZSM5 (x = 0 to 10 wt.%, SiO_2/Al_2O_3= 23, 55 and 280) catalysts was proposed in the present investigation. Characterization studies revealed the formation of different Zn species depending on the Zn loading. Furthermore, XPS analysis confirmed strong interaction of Zn species with the electronegative oxygen atom of zeolite framework. Zn deposition enhanced the selectivities of aromatics and BTEX. The Zn species presence suppressed the hydride transfer reaction and promoted the dehydrogenation reaction resulting in higher selectivities of aromatics and BTEX. The rising temperature and declining in WHSV enhanced the selectivities of aromatics and BTEX. The high pressure had adverse effect on the selectivities of aromatics. The maximum selectivity of total aromatics (74.83%) and BTEX (68.75%) were achieved at 723 K, 1 bar pressure and 0.75 h^{-1} of WHSV. The coke analysis revealed the formation of polynuclear aromatic coke at high pressure. The alkyl substitute aromatics and polynuclear aromatics were formed predominately on pure HZSM5 than 5Zn-HZSM5. A plausible reaction mechanism was suggested considering the products distribution.

Poster 10

Chemical Modification and Characterization of Cellulose Acetate

Abdulrahman Musa

Center for Refining & Advanced Chemicals, King Fahd University of Petroleum & Minerals, Dhahran, Saudi Arabia e-mail: absudais11@yahoo.com; abdulrahman.musa@kfupm.edu.sa

Abstract

During offshore oil extraction activities, and even in the onshore refining processes, contamination of water bodies is inevitable due to oil spillage. The problem has a direct negative consequence on our lives since we depend on portable water for good health. One of the contemporary challenges facing the oil and gas industries is the difficulty of adopting an effective technology and economically viable process for treatment of polluted water due to their industrial activities. Cellulose acetate (CA) is a polysaccharide used as an absorbent for the filtration of contaminated water. Because it can remove both organic and inorganic pollutants from wastewater due to its fibrous network structure. However, it has poor chemical and thermal stability that makes it unsuitable for sustainable water treatment. Here, we report the chemical modification of cellulose acetate using functionalized benzoxazine (FB) moiety to enhance the structurSal and thermal properties of the polymer.

Keywords: Cellulose acetate, Benzoxazine, Oil production, water treatment, Chemical modification

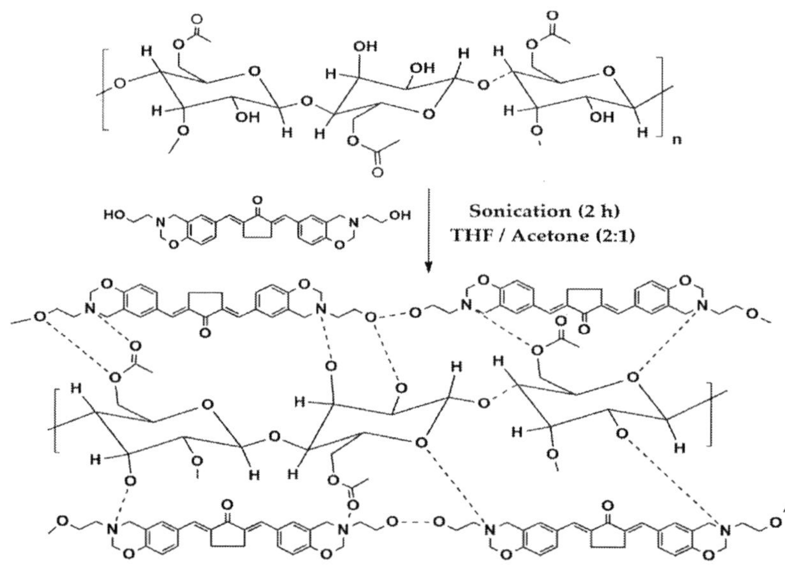

Scheme: The chemical modification of cellulose acetate using functionalized benzoxazine moiety

Poster 11

Role of Naphthenic-Aromatic Hydrocarbon in Autoxidation of Heavy Vacuum Gas Oil for Carbon Fiber Precursors

Mustafa M. Amin[a,b], Mohammad M. Hossain[a,b], Muhammad N. Siddiquee*[a,b]

[a] *Department of Chemical Engineering*

[b] *Center for Refining & Advanced Chemicals (IRC-RAC),*

King Fahd University of Petroleum & Minerals (KFUPM), Dhahran 31261, KSA

*Tel: +966-13-860-2353, *e-mail: muhammad.siddiquee@kfupm.edu.sa*

Abstract

Carbon fiber is an ultra-light but very strong material that is being increasingly used in different industries. The demand for carbon fiber has increased significantly from 26,500 tons in 2009 to 118,000 tons in 2022 resulting in a revenue boost of US$4.66 billion. Most currently used carbon fiber is PAN (polyacrylonitrile) based which is very expensive. Petroleum and coal tar pitches are potential options, but their softening point is low. They require pretreatment to improve the softening point, aromaticity, and viscosity for the melt-spinning steps and to improve the strength of the produced carbon fiber. In this regard, developing an economical and sustainable pretreatment process to produce carbon fiber precursors is highly desirable. The goal of the current study was to investigate the role of a naphthenic-aromatic solvent, tetralin, in developing the pretreatment process. Heavy vacuum gas oil (HVGO), a low-value heavy petroleum fraction, was used as a feed material. The methodology involved oxidation temperatures of 160 °C and 190 °C, a fixed oxidation time of 24 hours, and varying concentrations (0-30 wt %) of tetralin, which is known to form addition (heavy) products during autoxidation, and bubbling of air at a fixed value of 155 mL/min while maintaining mixing at 200 rpm. The oxidation of pure HVGO (0 wt % tetralin) showed a viscosity increase from 4643 cP to 6338 cP and 7411 cP, respectively at 160 °C and 190 °C for 24 h. The oxidation of 5 wt % tetralin in HVGO resulted 2.3 times viscosity increase at 160 °C and a 3.1 times viscosity increase at 190 °C for 24 h. Beyond 10 wt% tetralin in HVGO, the viscosity increase was insignificant for oxidation performed at identical conditions. It might be due to the dilution effect. It suggests that the use of a lower percentage of tetralin at higher temperatures would be beneficial to produce carbon fiber precursors from HVGO. Anticipated outcomes of the free radical addition include an increase in molecular weight, aromaticity, cross-linking, and softening point in the resulting product. This would be due to the formation of heavier products, the removal of lower boiling fractions, and ring closure. The hydrocarbon fractions treated in this manner hold potential as a precursor for carbon fiber.

Keywords: Carbon Fiber, Autoxidation, HVGO, Softening Point, Viscosity, Precursors.

Poster 12

Methanol Synthesis from CO_2 Hydrogenation Process: From Catalyst Design to Technology Development

Nagendra Kulal

Center for Refining & Advanced Chemicals (CRAC), King Fahd University of Petroleum & Minerals, Saudi Arabia

Abstract

Every year, energy consumption increases to meet countries' economic development. To fulfill this, fossil fuels are employed as energy sources. The combustion of natural gas, coal, oil, wood, and chemical processes greatly increases CO_2 concentrations in the atmosphere. CO_2 is a significant ingredient in greenhouse gases, and its concentration in the atmosphere is growing at an alarming rate. However, the world will continue to rely on fossil fuels in the next few decades. Researchers are attempting to mitigate CO_2 emissions through the direct capture of CO_2 from industry and its conversion. CO_2 hydrogenation to methanol is one of the promising routes for the utilization of CO_2. Methanol can be used as a precursor for the synthesis of different chemicals and is also an alternative fuel. In this study, we explain recent developments and remaining challenges of CO_2 utilization for the efficient production of methanol. Also includes novel technologies, approaches, and current barriers for the conversion of CO_2 to methanol through heterogeneous catalysis. Finally, development of new ideas for CO_2 hydrogenation to methanol on a large scale will be discussed.

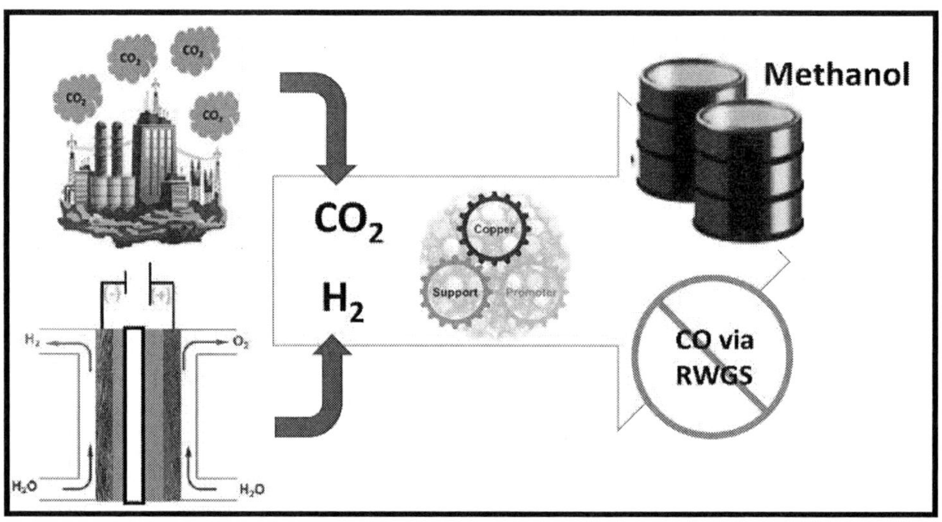

Figure : CO_2 hydrogenation to methanol

Poster 13

Well-designed glucose precursor carbon/g-C_3N_4 nanocomposite for enhanced photocatalytic CO_2 conversion to fuels

Abdullah Bafaqeer

Center for Refining & Advanced Chemicals, King Fahd University of Petroleum & Minerals, Saudi Arabia Email address: abdullah.bafaqeer@kfupm.edu.sa

Abstract

Fabrication of glucose precursor carbon-doped g-C_3N_4 nanocomposite (C/g-C_3N_4) for enhancing photocatalytic CO_2 reduction into syngas (CO, CH_4) has been investigated. The samples were successfully synthesized via a two-step thermal treatment and tested in a fixed bed reactor under visible light. The 0.2% glucose precursor carbon-doped over g-C_3N_4 photocatalyst has demonstrated excellent activity in converting CO_2 to CO and CH_4 under visible light. The main product yield, CO of 898.9 μmol g-cat^{-1} was produced over 0.2% C/g-C_3N_4, which is 4.6 folds the amount of CO obtained over the g-C_3N_4 (196.8 μmol g-cat^{-1}). The XPS results confirmed the formation of a C-O-C bond between carbon and g-C_3N_4, resulting in a strong interaction between carbon and g-C_3N_4. Carbon-doped g-C_3N_4 possesses a narrow energy band and the ability to effectively absorb solar light, which enables efficient transportation of electrons generated by photon excitation. Possible reaction mechanisms for photoreduction of CO_2 over carbon-doped g-C_3N_4 photocatalyst were proposed in order to understand the movement of electrons and holes. This work provides a simple method for designing highly efficient carbon-based photocatalysts for potential application in photocatalytic CO_2 reduction using solar energy.

Poster 14

TiO_2 based photocatalyst for solar hydrogen production from water

Muhammad Waqas[a], Saheed A Ganiyu[a,b], and Khalid Alhooshani[a,b]

[a] *Center for Refining & Advanced Chemicals, King Fahd University of Petroleum & Minerals, Dhahran, Saudi Arabia*

[b] *Department of Chemistry, King Fahd University of Petroleum & Minerals, Dhahran, Saudi Arabia*

Abstract

Constructing hollow multi-shelled structures has a significant effect on promoting light absorption property of catalysts and enhancing their performance in solar energy conversion applications. In this report, we will utilize a facile hydrothermal strategy to develop an innovative TiO_2 catalyst having multiple-shells and numerous pores windows for solar water splitting. The synthesis would be green in which the utilize starting materials are water, glucose and titanium salt. The unique hollow multi-shelled structure has following advantages: 1) increased the light absorption ability of the constructed photocatalysts and 2) the intricate shells boosted the collection efficiency of the photogenerated charge carriers, 3) The pore windows facilitate the diffusion of electrolytes at reactive surfaces to execute the photo-redox reaction.

Keywords: Hollow, Muti-shelled, Pore windows, Solar energy, Hydrogen

Poster 15

Highly efficient Ni/SiO_2-MgO catalyst for CO_2 methanation in synthetic natural gas production: Thermodynamics and catalytic insights

Ijaz Hussain[a], Saheed A. Ganiyu[a,b], Khalid Alhooshani[a,b]

[a] *Center for Refining & Advanced Chemicals, King Fahd University of Petroleum & Minerals, Dhahran, Saudi Arabia*

[b] *Department of Chemistry, King Fahd University of Petroleum & Minerals, Dhahran, Saudi Arabia*

Abstract

Climate change is a significant worry for the world due to carbon dioxide (CO_2) emissions from the excessive usage of fossil fuels. One of the most prestigious processes for converting CO_2 into value-added goods, including synthetic natural gas, is the CO_2 methanation process. However, developing an effective catalytic system remains a significant challenge. The present research focuses on thermodynamics and catalytic insights over a highly efficient Ni/SiO_2-MgO nanocomposite catalyst for effective CO_2 methanation using Ni/SiO_2-MgO nanocomposite catalyst for synthetic natural gas production. Ni/SiO_2-MgO nanocomposite catalyst was synthesized using the impregnation method. Different techniques such as XRD, TEM, SEM, N_2 adsorption analysis, and FTIR spectroscopy were used to characterize the synthesized catalysts. The CO_2 methanation reaction was performed, and it was found that the 5%Ni/SiO_2-MgO nanocomposite catalyst with a 1:1 ratio of SiO_2-MgO exhibited higher CH_4 selectivity (83%) and CO_2 conversion (90%) due to the synergistic effect of uniformed distributed Ni active phase on MgO and SiO_2. In addition, the synthesized 5%Ni/SiO_2-MgO nanocomposite catalyst exhibited high thermal stability and significantly suppressed coke deposition. This research offers a fundamental understanding of CO_2 methanation over Ni/SiO_2-MgO nanocomposite catalysts, which could be used to produce methane on a commercial scale.

Poster 16

Enhancing the efficiency of Ti_2C MXene electrocatalyst via zinc oxide nanorod intercalation for CO_2 electrochemical conversion to methane selectively

Abdulalhi Abdulhakam

Department of Chemistry, King Fahd University of Petroleum & Minerals, Dhahran, Saudi Arabia

Abstract

The industrial revolution has led to an increase in the burning of fossil fuels which has in turn skyrocketed the concentration of CO_2 in the atmosphere, a major cause of global warming and other environmental pollution. This study reports the use of a novel material (Ti_2C-ZnO nanocomposite) for CO_2 reduction via electrochemical technique to obtain valuable chemicals, which have proven to be a strategic way to mitigate the menace of CO_2 emission. Herein, MXene (Ti_2C), Zinc oxide nanoparticles, and Ti_2C-ZnO nanocomposites were synthesized via thermal treatment and etching of the "Al" layer from the MAX (Ti_2AlC) precursor, hydrothermal and wet impregnation methods respectively. Physical techniques like XRD, EDS, SEM, TEM, and electrochemical techniques like ECSA, TAFEL, EIS characterizations were utilized to gain valuable insights about the performance of the materials. The modified catalyst demonstrated a significant enhancement of the catalytic performance of the material. The Ti_2C- ZnO_5 nanocomposite, among others, showed the best efficiency and high selectivity to produce CH_4 at 96.6% faradaic efficiency with a low applied potential of -0.56 VRHE, while suppressing H_2 evolution greatly. It shows about 35.2% and 32.1% improvement compared to the pristine Ti2C and ZnO NPs respectively. This improved electrochemical activity may be linked to a constructive synergistic effect between the Ti2C and ZnO NPs, which promotes efficient charge transfer, increased active sites and high surface area for CO_2 reduction. Overall, this study presents a viable approach for CO_2RR electrochemically using Ti_2C- ZnO nanocomposites which could have significant implications to produce valuable feedstocks sustainably. Additionally, it provides an efficient and greener technique of methane production in contrast to the conventional methods which generate lots of greenhouse gases.

Keywords: CO_2 utilization, Electrochemical Reduction, Sustainability, Green methane production.

Poster 17

Dual activity of zinc oxide-MXene nanocomposite for enhancing the electrochemical conversion of CO_2 to Value-added Product

Abdulalhi Abdulhakam

Department of Chemistry, King Fahd University of Petroleum & Minerals, Dhahran, Saudi Arabia

Abstract

This study reports the use of a novel nanocomposite material composed of ZnO-MXene for CO_2 reduction via electrochemical technique which have proved to be a strategic and sustainable approach to obtain valuable feedstocks to curtail the global concern of skyrocketing concentration rise of CO_2 in the atmosphere which is detrimental to human health and weather/climate conditions. The ZnO nanorods, MXene (Ti_2C) and the nanocomposites were synthesized via hydrothermal, thermal treatment, and wet impregnation methods respectively. XRD, EDS, SEM, TEM, and FTIR characterization techniques were utilized to confirm the materials. Linear sweep voltammetry, cyclic voltammetry and controlled potential coulometry were used to evaluate the electrochemical performance of the nanocomposites, while the gaseous products were detected and quantified using gas chromatography. The ZnO–MX_{10} and ZnO–$MX_{2.5}$ nanocomposite showed a high selectivity towards the production of CH_4 and CO at 79.3% and 76.8% FE at an applied potential of -0.56 VRHE and -0.78 VRHE respectively, while suppressing H2 evolution greatly. This catalyst demonstrated a dual selectivity action for the electrochemical reduction of CO_2 at different potentials, which is shows an improved activity of about 24% for ZnO–MX_{10} and 13% for ZnO–$MX_{2.5}$ when compared to the bare ZnO catalyst. This improved electrochemical activity may be ascribed to a constructive cooperative effect between the ZnO and MXene components, which promotes efficient charge transfer, increased active sites and high surface area for CO_2 reduction. Overall, this study presents a viable approach for CO2RR electrochemically using ZnO-MXene nanocomposites which could have significant implications to produce valuable feedstocks sustainably.

Keywords: CO_2 utilization, Electrochemical reduction, Sustainability, Green methane, CO production.

Poster 18

Highly Efficient Fibrous Silica Lanthanum Oxide-Supported Nickel Catalyst for Dry Reforming of MethaneMohammed Mosaad Awad, Ijaz Hussain, Saheed A. Ganiyu, Khalid Alhooshani

Department of Chemistry, King Fahd University of Petroleum & Minerals, Dhahran, Saudi Arabia

Abstract

The current study focuses on the dry reformation of methane (DRM) reaction, which converts methane (CH_4) and carbon dioxide (CO_2) into useful syngas using a Ni-loaded Fibrous Silica Lanthanum Oxide (Ni/FSL) catalyst. The Ni/FSL catalysts with varying nickel loadings were synthesized using a wet impregnation method. The synthesized catalysts were characterized by different techniques such as Transmission Electron Microscopy (TEM), Energy-Dispersive X-ray Spectroscopy (EDS), Scanning Electron Microscopy (SEM), X-ray diffraction (XRD), N_2-adsorption-desorption analysis, Thermogravimetric analysis (TGA), Fourier-transform infrared spectroscopy (FTIR), and Raman spectroscopy. The catalytic performance of Ni/FSL catalysts was performed using a fixed-bed reactor at atmospheric pressure. It was found that the Ni/FSL (20% Ni loading) catalyst of unique fibrous morphology exhibited superior catalytic preperformance towards CO_2 conversion (90%) and CH_4 conversion (88%) with a 1.68 H2/CO ratio at the temperature of 750 ^{o}C. Additionally, the Ni/FSL catalyst also showed remarkable stability and resistance to carbon deposition, showing high thermal stability under the same reaction conditions. The synergistic effects of Ni/FSL, which aided the activation of CH_4 and CO_2, are credited with the observed catalytic performance, which led to increased syngas generation. It is concluded that the 20Ni/FSL catalyst has great potential to be applied on an industrial scale for syngas production via DRM, mitigating greenhouse gas emissions.

Poster 19

Controlled Synthesis of Zinc Layered Double Hydroxides for Superior Electrochemical CO_2 Reduction

Omer Ahmed Taialla[a], Umar Mustapha[a,b], Khalid Omer[a], Saheed A. Ganiyu[a,b] and Khalid Alhooshani[a,b,*].

[a]*Department of Chemistry, King Fahd University of Petroleum and Minerals, Dhahran, Saudi Arabia*

[b]*Interdisciplinary Research Center for Refining & Advanced Chemicals, King Fahd University of Petroleum & Minerals, Dhahran, Saudi Arabia*

Abstract

The increase in anthropogenic carbon dioxide emissions has significantly and detrimentally affected our ecosystems. Electrochemical reduction (ECR) of CO_2 is one of the emerging alternatives to mitigate atmospheric CO_2 concentration and achieve a carbon-neutral state, primarily due to its capability to utilize renewable sources efficiently. However, the CO_2-ECR currently suffers from low efficiency and inadequate selectivity. Therefore, there is a pressing need for an efficient electrocatalyst that can enhance desired product selectivity and improve the process's efficiency. In this study, the electrocatalytic CO_2-ECR activity of Zn-layered double hydroxides incorporated chlorine (ZnLDH-Cl) was studied. ZnLDH-Cl catalysts were prepared, and CO_2 RR was performed using CO_2-saturated $KHCO_3$ electrolyte to evaluate the electrocatalytic ability of ZnLDH-Cl catalysts. ZnLDHs intercalated with different species (Ca^{+2}, Na^+ and NH) were synthesized using metal chloride through hydrolysis process. The synthesized ZnLDH-Cl catalysts exhibited the electrocatalytic CO_2-ECR activity for CO formation. The highest faradaic efficiency was achieved with ZnLDH-Cl/Ca demonstrating 99.7% at -0.96 V vs RHE at remarkably low overpotential. In addition, the catalyst maintains high stability reach up to 20.8h.

Poster 20

Spent FCC catalyst in pesticides microextraction: a sustainable approach towards waste recycling/reuse

Shaima' Ibrahim Alsabbahen[1], Abdulkadir Tanimu[2], Shehzada Muhammad Sajid Jillani*[3], Khalid Alhooshani*[1, 2]

[1]*Department of Chemistry, King Fahd University of Petroleum & Minerals, Dhahran, Saudi Arabia.*

[2]*Center for Refining & Advanced Chemicals, King Fahd University of Petroleum & Minerals, Dhahran, Saudi Arabia*

[3]*IRC Membranes and Water Security, King Fahd University of Petroleum & Minerals, Dhahran, Saudi Arabia*

Abstract

Anthropogenic activities are introducing and increasing pollution in the environment. To achieve high agricultural yield with healthy crops, extensive use of pesticides, insecticides and fertilizers is in fashion. Such pesticides and their residues in water bodies are hazardous and poisonous for human health. The lower concentrations of such pollutants make it difficult to analyze using normal gas chromatographic techniques. Therefore, a sample preparation step is included to isolate and preconcentrate the analyte of interest. There are many extractions and microextraction methods reported in literature with their intrinsic advantages and disadvantages. The membrane supported micro-SPE is considered as more environmentally friendly and effective. In this work, we utilized spent zeolitic catalyst derived from a local FCC refinery as sorbent for the microextraction of pesticides in water samples. The spent catalysts are waste materials that are often disposed in landfills and this is environmentally unfriendly because the spent catalyst flies in the atmosphere and when inhaled, it accumulates in the respiratory track and lungs, which is detrimental to human health. The performance of spent catalyst material was not only compared with regenerated or pristine zeolite but also a modified version of it was utilized after functionalization. The materials were packed inside the membrane bag making it a micro-SPE tool. The spent zeolitic catalyst performed better for the extraction of six pesticide residues. The materials were characterized powder XRD, SEM-EDX, TGA, BET and FTIR. All analytical parameters were evaluated for method validation including linearity, correlation coefficient, LOD, LOQ, and reproducibility. The wastewater samples were tested using the developed method. The sample was also spiked, and relative recoveries were also evaluated. The developed method has proved sensitive and reproducible for the determination of pesticide residues on wastewater analysis.

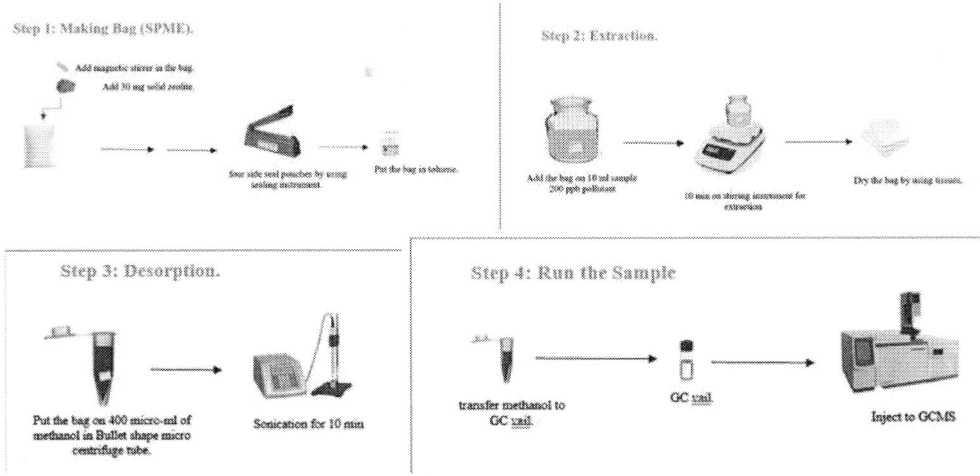

Figure 1: Typical Presentation of micro-SPE procedure.

Poster 21

Turning Waste into Value: K- Promoted Red Mud as an Effective Catalyst for CO_2 Hydrogenation to Olefins

Mahbuba Aktary[a], Huda S. Alghamdi,[b] Mohammed Sanhoob,[a] Atif S. Alzharani,[b]* M. Nasiruzzaman Shaikh[a]

[a]*Materials Science and Engineering Department (MSE), King Fahd University of Petroleum and Minerals (KFUPM), Dhahran, Saudi Arabia*

[b]*Interdisciplinary Research Center for Hydrogen and Energy Storage (IRC-HES), King Fahd University of Petroleum & Minerals, Dhahran 31261, Saudi Arabia*

e-mail: atif@kfupm.edu.sa (Atif S. AlZaharani)

Abstract

Reducing the CO_2 emissions is an urgent global priority. In this context, several mitigation strategies, such as CO_2 tax, stringent legislation, etc., are adopted to stop the further deterioration of the earth's environment. The quest for inventive catalysts in CO_2 hydrogenation has been driven by the growing demand for eco-friendly and sustainable practices within the chemical industry for green chemical production such as olefins. Red mud (RM), due to its unique blend of diverse metal oxides, is emerging as a potential catalyst for this purpose. This abstract roots into the potential breakthroughs associated with the application of K-promoted RM catalysts in the context of CO_2 hydrogenation for olefin production. Various weight percentages of K were introduced to enhance the active sites of RM. Fascinatingly, both pure RM and HCl-treated RM exhibited comparable performance in CO_2 conversion, yielding 21.5% and 22.1%, respectively. Notably, the selectivity for olefins was substantially higher in the case of HCl-treated RM (12.1%) compared to untreated red mud (7%) at elevated temperatures of 375°C and a pressure of 30 bar. Among different weight percentages of K-promoted RM (1%, 2%, 3%, and 5%), the 3% variant exhibited superior performance. It has achieved the highest conversion rate at 27.5% along with an impressive olefin selectivity of 26.7%. Interestingly, an excess of K in the RM catalyst adversely affected CO_2 conversion, as demonstrated by the 5% K-promoted RM variant, which, while offering the highest conversion rate, saw a decline in olefin selectivity. This experimental research uncovers crucial insights, mechanisms, and practical applications of K-promoted RM catalysts, shedding light on their pivotal role in reshaping olefin production and advancing the sustainability goals of the chemical industry.

Keywords: Carbon dioxide, Catalyst; Hydrogenation; Olefins; Red mud

Poster 22

Production of Dimethyl Carbonate from CO_2 by using Cerium Oxide-Based Catalyst

Mohammed Alqarni, Samiyah Aljendan, Niladri Maity

Interdisciplinary Research Center for Refining & Advanced Chemicals, King Fahd University of Petroleum & Minerals, Dhahran-31261, Saudi Arabia

Abstract

Catalytic conversion of carbon dioxide (CO_2) into value-added chemicals is one of the pressing challenges for the scientific community to mitigate CO_2 emissions into the atmosphere after the burning of fossil fuels. In fact, the industrial-scale chemical utilization of CO_2 is presently limited to the production of bulk chemicals, such as urea, salicylic acid, cyclic carbonates, and polypropylene carbonate. The global market of dimethyl carbonate (DMC) is forecasted to reach $2.06 billion market value by 2030 at a compound annual growth rate (CAGR) of 7.1 % from $1.19 billion in 2022 due to its wide application as feedstock for manufacturing polycarbonates, electrolyte in lithium-ion batteries, low-volatile green solvent, chemical intermediate for polyurethane, and potential use as a fuel additive for diesel. DMC is produced in the chemical industries using highly toxic and expensive starting materials, such as phosgene ($COCl_2$), carbon monoxide (CO), dialkyl sulfate ($(OR)_2SO_4$), and epoxides under hazardous synthesis routes.

The direct production of DMC from CO_2 and methanol (CH_3OH) is an economically viable and green synthesis process owing to the high abundance and nontoxicity of starting materials generating water (H_2O) as a side product (**Scheme 1**).

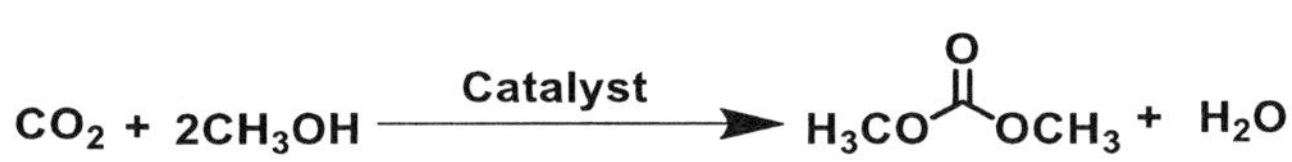

Scheme 1: Catalytic conversion of CO_2 into dimethyl carbonate.

The catalyst systems mostly explored in this aspect are based on ionic liquids, alkali carbonates, heteropoly acids, supported metal nanoparticles, and metal oxide catalysts. Among those, metal oxide catalysts have emerged as the dominant choice by exhibiting remarkable resistance to deactivation through oxidation, and their stability in air. Among the metal oxide catalysts, the amphoteric cerium oxide (CeO_2) based materials with surface oxygen vacancy are highly promising due to their excellent efficiency in CO_2 activation and promoting the carboxylation of CH_3OH to result in DMC [1-3]. However, their tendency towards the low yield of DMC can be mitigated by the addition of an efficient dehydrating reagent. A variety of CeO_2-based novel catalyst materials synthesized in this aspect show high efficiency towards the yield of DMC at moderate reaction conditions.

References

[1] Stoian, D.; Medina, F.; Urakawa, A. *ACS Catal.* **2018**, *8*, 3181–3193.
[2] Tomishige, K.; Gu, Y.; Nakagawa, Y.; Tamura, M. *Front. Energy Res.* **2020**, *8*, 117 (1-24).
[3] He, Z.-H.; Sun, Y.-C.; Wei, Y.-Y.; Wang, K.; Wang, W.; Chen, Z.; Z.-Y. Wang, Tian, Y.; Liu, Z.-T. *Fuel*, **2022**, *325*, 124945 (1-11).

Poster 23

Molecular Dynamic Simulation of Refrigerant Separation Using Zeolite: A Comprehensive Study on Adsorption

Abrar A. Elhussein[1] and Wael A. Fouad[1,2]

[1]Chemical Engineering Department; [2] Center for refining & Advanced Chemicals, King Fahd University of Petroleum & Minerals, Dhahran, Saudi Arabia

Abstract

In recent decades, heightened concerns about global warming underscore the need for sustainable practices. The widespread use of fluorocarbon derivatives in industries contributes significantly to global warming. Recycling and reusing fluorocarbons are crucial for curbing environmental emissions and meeting future demands, especially with the rapid phase-out of certain fluorocarbons. Additionally, improving adsorption/ desorption processes for hydrofluorocarbons (HFCs) holds significant implications for various heat transformation applications. The recycling process, achieved through a separation process, is integral to addressing this environmental concern. The adsorption process using zeolite emerges as a cost-effective and more efficient alternative to traditional distillation methods. This dual focus on environmental responsibility and technological innovation positions zeolite-based adsorption as a promising solution for sustainable refrigerant separation and heat transformation applications.

This study explores the potential of Zeolite as advanced adsorbent materials, leveraging their high surface area, porosity, stability, and tunability.

Our approach involves employing theoretical modeling through Molecular Dynamics (MD) simulations to modify the adsorbate structure of materials like zeolite. This modification aims to create a diverse range of options for adsorption, particularly focusing on a wide array of refrigerants, including various HFCs. The investigation scrutinizes the influence of fluorine atoms in the adsorption process, shedding light on their impact on the efficiency and selectivity of the adsorbents.

Poster 24

Ruthenium supported on lanthanide oxyhydrides: as an efficient ammonia synthesis catalysts

Walid Al Maksoud;[a] Rohit Rai;[a] Yu Cao;[a, b] Yuki Sasahara;[b] Hiroaki Sai;[b] Omar El Tal;[c] Mohamed N. Hedhili;[d] Yoji Kobayashi[a*].

[a] *Chemistry Program, KAUST Catalysis Center, Division of Physical Science and Engineering, King Abdullah University of Science and Technology, Thuwal, 23955-6900 Saudi Arabia.*

[b] *Department of Energy and Hydrocarbon Chemistry Graduate School of Engineering, Kyoto University Nishikyo-ku, Kyoto 615-8510, Japan.*

[c] *Analytical Core Labs, King Abdullah University of Science and Technology, Thuwal, 23955-6900 Saudi Arabia.*

[d] *Imaging and Characterization Department, Core Labs, King Abdullah University of Science and Technology, Thuwal, 23955-6900 Saudi Arabia.*

**Corresponding Author Prof. Yoji Kobayashi; Contact email: yoji.Kobayashi@kaust.edu.sa*

Abstract

Ammonia is considered among the most important processes in chemical industries due to its large application such as fertilizer, nitrogen-based chemicals, and others. Recently, NH_3 has also attracted great attention as an excellent candidate for hydrogen storage and transportation due to its high volumetric and gravimetric hydrogen capacity and facile liquefaction under mild conditions. In addition, CO_2-free hydrogen is available from NH_3 decomposition.[1-3]

In the last decade, new materials incorporating electrides, nitrides, and hydrides have been discovered, which show high activity for ammonia synthesis.[4] Among these materials, oxyhydrides were also appeared and used as an active support for ammonia synthesis.[5, 6]

In this work, we described the lanthanide oxyhydrides as promising catalysts for ammonia synthesis reactions owing to their unique feature and hydride liability. The newly prepared lanthanide oxyhydride materials are expected to impart additional basicity as compared to oxides for enhanced ammonia synthesis catalysts.

After the successful preparation of lanthanide oxyhydrides, their basic properties have been intensively investigated through several methods such us CO_2-TPD, *in-situ* CO_2-DRIFTS, and for the first time by Hammett Indicators. Furthermore, small RuNPs supported on the oxyhydrides (GdOH, SmOH, PrOH, and LaOH) were successfully prepared and exhibited excellent activity in ammonia synthesis reaction without adding promoters. Moreover, the kinetics parameters demonstrated the absence of hydrogen poisoning, and a low energy activation was observed contrary to the conventional Ru-based catalysts. Further characterizations were applied to highlight the difference in activity between RuNPs supported on lanthanide oxyhydrides and oxides (*in-situ* CO-DRIFTS and XPS analyses), and we found that electron donation effects can occur from oxyhydrides to Ru.

1. Q. Wang, J. Guo and P. Chen, *Journal of Energy Chemistry*, 2019, **36**, 25-36.
2. C. Smith, A. K. Hill and L. Torrente-Murciano, *Energy & Environmental Science*, 2020, **13**, 331-344.
3. N. Morlanés, S. P. Katikaneni, S. N. Paglieri, A. Harale, B. Solami, S. M. Sarathy and J. Gascon, *Chemical Engineering Journal*, 2021, **408**, 127310.
4. H. Hosono and M. Kitano, *Chemical Reviews*, 2021, **121**, 3121-3185.
5. Y. Tang, Y. Kobayashi, N. Masuda, Y. Uchida, H. Okamoto, T. Kageyama, S. Hosokawa, F. Loyer, K. Mitsuhara, K. Yamanaka, Y. Tamenori, C. Tassel, T. Yamamoto, T. Tanaka and H. Kageyama, *Advanced Energy Materials*, 2018, **8**, 1801772.
6. H. Yamashita, T. Broux, Y. Kobayashi, F. Takeiri, H. Ubukata, T. Zhu, M. A. Hayward, K. Fujii, M. Yashima, K. Shitara, A. Kuwabara, T. Murakami and H. Kageyama, *Journal of the American Chemical Society*, 2018, **140**, 11170-11173.

Poster 25

Fibrous Silica Supported Bi/Ni Oxide Catalysts for Oxidative Dehydrogenation of n-Butane

Ridhwan Lawal [1], Hassan Alasiri [1,2], Gazali Tanimu [2], Kulal Negandra[2], Ziyauddin Qureshi

[1]*Department of Chemical Engineering, King Fahd University of Petroleum & Minerals, Dhahran 31261, Saudi Arabia.*

[2]*Center for Refining and Advanced Chemicals, King Fahd University of Petroleum & Minerals, Dhahran 31261, Saudi Arabia.*

ABSTRACT

Oxidative dehydrogenation of n-butane to produce 1,3-butadiene was investigated over a series of catalysts having Ni-Bi percentage weight loading of 10-10, 10-15, 15-10, 20-10, 20-20 and 20-30 dispersed over fibrous silica support. Conducted at atmospheric pressure, the experiments employed a fixed-bed reactor at temperatures of 400°C, 450°C, and 500°C, with butane to oxygen feed ratios of 1, 2, and 4. Several characterization techniques, including XRD, BET, SEM, and TEM were utilized to analyze catalyst properties. The study revealed a correlation between the formation of NiO and Bi_2SiO_5 phases, influencing dehydrogenation activity and butadiene selectivity, respectively. Notably, the 20Ni-10Bi/FS catalyst exhibited superior performance, giving the highest butadiene yield of 20% at 500°C with a feed ratio of 1.

32nd Annual Saudi-Japan Symposium – 2023

Technology in Fuels & Petrochemicals

Innovative Catalyst Development

Venue: KFUPM Dhahran Techno Valley, Innovation Cluster Bldg., Auditorium 1-111

December 5-6, 2023

Day One: Tuesday, December 5, 2023

OPENING REMARKS SESSION — ***Chairman: Dr. Wael Fouad, KFUPM***

8:30 Opening Remarks

- ***Dr. Muhammad M. Al-Saggaf,*** *President, KFUPM*
- ***Mr. Hitoshi Masuda,*** *Special Counselor, JCCP*
- ***Dr. Ali A. Al-Meshari,*** *Senior Vice President, TOC, Saudi Aramco*
- ***Dr. Atsushi Muramatsu,*** *President, JPI*

SESSION ONE PETROLEUM CONVERSION — ***Chairman: Dr. Mohammad Al-Abdullah, Saudi Aramco***

9:00 1. **Keynote: JPI direction along carbon neutrality and introduction of our research, direct methane conversion over zeolite catalysts;** *Dr. Atsushi Muramatsu, President of JPI, Deputy Director (Soft Materials), Tohoku University, Japan*

9:30 2. **Atmospheric impact of fugitive CH_4 and H_2;** *Dr. Theis Solling, Center for Integrated Petroleum Research, CIPR, KFUPM*

9:50 3. **Crude to chemicals – old challenges and new opportunities;** *Dr. Lujain R. Alfilfil, Saudi Aramco R&DC*

10:10 Coffee break

SESSION TWO AI AND ML APPLICATIONS — ***Chairman: Dr. Hassan Aljama, Saudi Aramco***

10:40 4. **Application of ML to predict performance of oxidative dehydrogenation catalysts;** *Dr. Gazali Tanimu, Center for Refining & Advanced Chemicals (CRAC), KFUPM*

11:00 5. **Accelerating catalyst discovery using extrapolative ML approach;** *Dr. Takashi Toyao, Institute for Catalysis, Hokkaido University, Japan*

11:20 6. **AI and data-driven optimization of one-step crude to chemicals process;** *Ms Noor Sulais, R&DC, Saudi Aramco*

11:40 Prayer & Lunch Break

SESSION THREE PETROCHEMICALS-1 — ***Chairman: Dr. Zuhair Malaibari, KFUPM***

13:00 7. **Distillate hydrocracking catalyst and process to produce isomerate and steam feedstock;** *Dr. Ashok K. Punetha, Saudi Aramco*

13:20 8. **Molecular kinetic model development and parameter estimation for naphtha reforming;** *Dr. Syed A. Ali, CRAC, KFUPM*

POSTER SESSION *Coordinator: Dr. Rajesh Theravalappil, KFUPM*

13:45 P1. **2-Pentene cracking over bifunctional MFI-type zeolites;** *Mr. Mohammed Alkhunaizi, R&DC, Saudi Aramco*

13:50 P2. **Post-Synthesis Functionalization of Covalent Organic Frameworks for Carbon Dioxide Capture from Air (DAC);** *Mrs. Mona Al-Otaibi, R&DC, Saudi Aramco*

13:55 P3. **Resilience of transalkylation catalyst towards industrial contaminants;** *Mr. Mosab T. Kheyami, R&DC, Saudi Aramco*

14:00 P4. **Polypropylene cracking proceeding in micropores of MFI type zeolite;** *Mr. Tomohiro Fukumasa, Center Research Green Sustainable Chemistry, Tottori University*

14:05 P5. **Towards sustainable CO_2 valorization: Harnessing Cu single atoms and nanoparticles;** *Ms. Esraa Kotob, CHEM, KFUPM*

14:10 P6. **Catalytic conversion of LDPE plastic via pyrolysis process;** *Mr. Feras Alqudayri, CRAC, KFUPM*

14:15 P7. **Thermocatalytic pyrolysis of microalgae biomass over Y-zeolite catalyst towards clean fuel and valuable chemicals;** *Ms. Hayat A. Haddad, CHE, KFUPM*

14:20 P8. **Effect of plastic composition on synergetic interactions, kinetic & thermodynamic properties from co-pyrolysis of date palm waste and waste foam;** *Dr. Ahmad Nawaz, CRAC, KFUPM*

14:25 P9. **Enhanced selectivity of benzene-toluene-ethyl benzene and xylene in direct conversion of n-butanol to aromatics over Zn-HZSM-5 catalysts;** *Dr. Tatinaidu Kella, CRAC, KFUPM*

14:30 P10. **Chemical modification/characterization of cellulose acetate;** *Dr. Abdulrahman Musa, CRAC, KFUPM*

14:35 P11. **Role of naphthenic-aromatic hydrocarbon in autoxidation of heavy vacuum gas oil for carbon fiber precursors;** *Mr. Mustafa M. Amin, CHE, KFUPM*

14:40 P12. **Methanol synthesis from CO_2 hydrogenation process: from catalyst design to technology Development;** *Dr. Nagendra Kulal, CRAC, KFUPM*

14:45 P13. **Well-designed glucose precursor carbon/g-C3N4 nanocomposite for enhanced photocatalytic CO_2 conversion to fuels;** *Dr. Abdullah Bafaqeer, CRAC, KFUPM*

14:50 P14. **TiO_2 based photocatalyst for solar hydrogen production from water;** *Dr. Muhammad Waqas, CRAC, KFUPM*

14:55 P15. **Highly efficient Ni/SiO_2-MgO catalyst for CO_2 methanation in synthetic natural gas production: Thermodynamics and catalytic insights;** *Dr. Ijaz Hussain, CRAC, KFUPM*

15:00 P16. **Enhancing the efficiency of Ti_2C MXene electrocatalyst via zinc oxide nanorod intercalation for CO_2 electrochemical conversion to methane selectively;** *Mr. Abdulalhi Abdulhakam, CHEM, KFUPM*

15:05 P17. **Dual activity of zinc oxide-MXene nanocomposite for enhancing the electrochemical conversion of CO_2 to Value-added Product;** *Mr. Abdulalhi Abdulhakam, CHEM, KFUPM*

15:10 P18. **Highly efficient fibrous silica lanthanum oxide-supported nickel catalyst for dry reforming of methane;** *Mr. Mohammed Awad, CHEM, KFUPM*

15:15 P19. **Controlled synthesis of zinc layered double hydroxides for superior electrochemical CO_2 reduction;** *Mr. Omer Taialla, CHEM, KFUPM*

15:20 P20. **Spent FCC catalyst in pesticides microextraction a sustainable approach towards waste recycling/reuse;** *Ms Shaima' Alsabbahen, CHEM, KFUPM*

15:25 P21. **Turning waste in to value: K-promoted red mud as an effective catalyst for CO_2 hydrogenation to olefins;** *Ms Mahbuba Aktary, MSE/IRC-HES, KFUPM*

15:30 P22. **Production of dimethyl carbonate from CO_2 by using cerium oxide-based catalyst;** *Mr. Mohammed Alqarni, CRAC, KFUPM*

15:35 P23. **Molecular dynamics simulation of refrigerant separation using zeolite: A comprehensive study on adsorption;** *Ms. Abrar A. Elhussein, CHE, KFUPM*

15:40 P24. **Ruthenium supported on lanthanide oxyhydrides: as an efficient ammonia synthesis catalysts;** *Dr. Walid Al Maksoud, KAUST Catalysis Center*

15:45 P25. **Fibrous silica supported Bi/Ni oxide catalysts for oxidative dehydrogenation of n-butane;** *Mr. Ridhwan Lawal, CHE, KFUPM*

15:50 Poster Session Ends

32nd Annual Saudi-Japan Symposium – 2023

aramco

Technology in Fuels & Petrochemicals

Innovative Catalyst Development

Day Two: Wednesday, December 6, 2023

SESSION FOUR PETROCHEMICALS-2 — ***Chairman: Dr. Takashi Toyao, Hokkaido University***

8:30	**9.**	**Effect of gallium and platinum distribution encapsulated in silicalite-1 (MFI) zeolite on controlled propane dehydrogenation reaction;** *Mr. Fadhil A. Almukhtar, Saudi Aramco R&DC*
8:50	**10.**	**Highly efficient dehydrogenation of isopentane to isoprene: Selectivity control of the catalytic reaction field by electric internal heating system;** *Dr. Ryo Watanabe, Shizuoka University, Japan*
9:10	**11.**	**Insights on CO_2-mediated oxidative dehydrogenation of propane: Aspen plus simulation and in situ DRIFT experiment;** *Dr. Yahya Gambo, CRAC, KFUPM*
9:30	**12.**	**Light olefins cracking by zeolites prepared from refinery waste;** *Mr. Mohammad Rebh, Saudi Aramco R&DC*
9:50		***Coffee Break***

SESSION FIVE SUSTAINABILITY-1 — ***Chairman: Dr. Saheed Ganiyu, KFUPM***

10:20	**13.**	**CO_2 hydrogenation to carbon-neutral liquid fuels by powerful catalyst;** *Dr. Noritatsu Tsubaki, University of Toyama, Japan (Online Lecturer)*
10:40	**14.**	**CO_2 hydrogenation to lower olefins using iron supported catalysts;** *Dr. M. Nasiruzzaman Shaikh, Center for Hydrogen & Energy Storage, KFUPM*
11:00	**15.**	**Insights into plastic pyrolysis and bio-oil upgrading: process optimization and techno-economic analysis;** *Dr. Omar Abdelaziz, CHE, KFUPM*
11:20	**16.**	**Selective aromatics recovery by catalytic conversion of pyrolysis gas from carbon fiber reinforced plastic;** *Dr. Kazumasa Oshima, Kyushu University, Japan*
11:40	**17.**	**Challenges and opportunities in converting waste plastic into value-added products;** *Dr. Mohammad Nahid Siddiqui, CHEM, KFUPM*
12:00		***Prayer & Lunch Break***

SESSION SIX SUSTAINABILITY-2 — ***Chairman: Dr. Kazumasa Oshima, Kyushu University***

13:00	**18.**	**Challenge of green hydrogen production from air by direct air electrolysis;** *Dr. Etsushi Tsuji, Tottori University, Japan (Online Lecturer)*
13:20	**19.**	**Adjusting the crude oil–to–chemical process using unconventional reactors and catalyst Formulations,** *Dr. Pedro Castano, KAUST Catalysis Center, KAUST*
13:40	**20.**	**Fueling future aviation with CO_2 to jet technology;** *Mr. Arthur Foutsitzis, Honeywell UOP, KSA*
14:00	**21.**	**3D porous polymers for selective removal of CO_2 and H_2 storage;** *Dr. Othman Al Hamouz, Chemistry Department, KFUPM*
14:20	**22.**	**Metal-organic frameworks functionalization and design strategies for CO_2 capture;** *Ms. Nawal M. Alghoraibi, R&DC, Saudi Aramco*
14:40		***Closing Remarks, Symposium Ends*** *Dr. Wael Fouad, CRAC, KFUPM*

Each presentation includes 5-minutes Q&A

Symposium supported by: KIKX